The Fastest Gun in the Big Red One
1st Battalion 7th U.S. Artillery in Vietnam
1965 - 1970

This book is dedicated to the men of Battery A, 1st Battalion, 7th Artillery, known during the Vietnam War as the "Pheons" of the Big Red One. And, it is further dedicated to the names carved on "The Wall".

"The Big Red One" translates into Vietnamese as "Big Red Brothers"...the Vietnamese refer to the division as "...a unit that won a hundred battles in a hundred fights."

-David Maraniss
Author of
They Marched into Sunlight

The Fastest Gun in the Big Red One
1ˢᵗ Battalion 7ᵗʰ U.S. Artillery in Vietnam
1965 - 1970

Other books by Walt Cross include:

Boomer Soldier
A Hot War – Cold War Story

Custer's Lost Officer; the Search for Lieutenant
Henry Moore Harrington, 7ᵗʰ U.S. Cavalry

From Little Big Horn to the Potomac;
The Story of Army Surgeon Dr. Robert W. Shufeldt

From the Beaches to the Baltic; the Story of the 7ᵗʰ
Armored Division in WWII
(editor)

Out West with Custer and Crook; the Story of
Colonel Verling K. Hart of the 7ᵗʰ and 5ᵗʰ U.S.
Cavalry

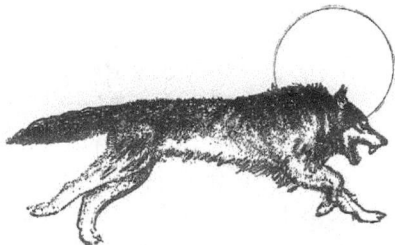

Dire Wolf Books
Copyright by Walt Cross © 2012
502 E. Liberty Avenue
Stillwater, Oklahoma 74075

Online at www.lulu.com/greenpheon7

First Edition

ISBN-13 978-0-9771926-9-4

Cataloging Data

Cross, Walt
The Fastest Gun in the Big Red One 1st Battalion 7th
U.S. Artillery in Vietnam 1965 - 1970
United States History – Vietnam 2. United States
Army – Vietnam 3. 1st Battalion, 7th U.S. Artillery.
I. Title

Front cover designed and painted by Walt Cross

MANUFACTURED IN THE UNITED STATES
OF AMERICA

Foreword

On the 23rd of June 1965, Private First Class Gerald Worster arrived by plane in Saigon, Republic of Vietnam. Not only was Gerald the first soldier to arrive from the 1st Infantry Division, known as the "Big Red One", he was also the first member of the 1st Battalion, 7th Artillery to set foot on Vietnamese soil. Even as he climbed from the plane, his comrades were loading aboard the USNS *Gordon* for the seaborne trip to Vietnam.

On 12 July, Charlie Battery of the 1st Battalion, 7th Artillery made landfall at Cam Ranh Bay, part of a taskforce with the 1st Battalion, 18th Infantry, 1st Infantry Division. This force was the first tactical US Army unit to be deployed in the Republic of Vietnam directly from the continental United States. It would not be the last.

On March 17, 1970 the 1st Battalion, 7th Artillery, 1st Infantry Division left Vietnam. During the last year of service men and officers of the battalion earned 1 Silver Star Medal[1], 3 Distinguished Flying Cross Medals, 15 Bronze Star Medals for Valor[2], 11 Army Commendation Medals for Valor, and 2 Vietnamese Honor Medals. All these redleg soldiers gave of themselves to the nation and some gave the last full measure of their devotion.

---

[1] This soldier was a member of Battery A, 1/7th Artillery.
[2] Two Bronze Star Medals for Valor and an Army Commendation Medal for Valor were awarded to members of Battery A, 1/7th Artillery.

Introduction

Before the 1st Battalion, 7th U.S. Artillery was created as an organic part of the First Infantry Division there was a similarly designated unit, the *1st Battalion, 7th U.S. Field Artillery*. The 1/7th Artillery was created, trained, and equipped to fight the Vietnam War.

Soon after the end of their Vietnam tour of duty the unit was deactivated and once again the *1/7th Field Artillery* would return to the Big Red One. They would list among the unit campaigns those earned by the 1/7th Artillery in Vietnam, although the 7th Artillery is not the parent unit of the 1/7th Field Artillery. A second battalion, this one an air defense unit, would also claim the battalion's Vietnam campaigns as their own. Neither of these units are descended from the 7th Artillery.

In point of fact the 7th U.S. Artillery is an orphan, an orphan of the Vietnam War with neither a parent unit nor a descendant unit. One of my motivations for penning this book is to preserve as much of the story of this fighting battalion as I can find.

During the Vietnam War the headquarters of division artillery instituted an award program that recognized the exceptional performance of individual artillery batteries. This award was dubbed "The Fastest Gun in the Big Red One". My battery first sergeant told me that Battery A of the battalion, the battery I served with, took that award eleven of the twelve months plus that I was assigned to the unit. And that is where the title of this book originates.

Acknowledgements

I wish to acknowledge and thank Mr. Bob Abbott, former sergeant of Battery A, 1st Battalion, 7th Artillery for providing important information regarding 7th Artillery operations in Vietnam.

Bob was section chief in the FDC (fire direction and control center) during and after the battle for Fire Support Base Jim. Bob would later serve as artillery liaison NCO with the 2nd Battalion, 16th Infantry of the Big Red One operating out of FSB Dominate. Bob would have one last contact with Alpha Battery at Fire Support Base Oklahoma before he rotated back to "the world" with the battalion colors.

Bob contacted me in 2003 regarding making a supporting statement for his application for VA support. I was happy to write him a statement and he received the support he and others earned as "Pheons" in Vietnam. Thanks Bob.

I'd also like to thank former 7th Artillery soldiers Russell Gallegos, Paul Jones, and Richard "Rick" Morrow for material support that assisted me in the writing of this book. Russ provided many photographs from his tour February 1968 to February 1969.

Walt Cross
Master Sergeant
U.S. Army (Retired)
Stillwater, OK 74075
June 14, 2012 the 237th birthday of the United States Army, hooah!

Table of Contents

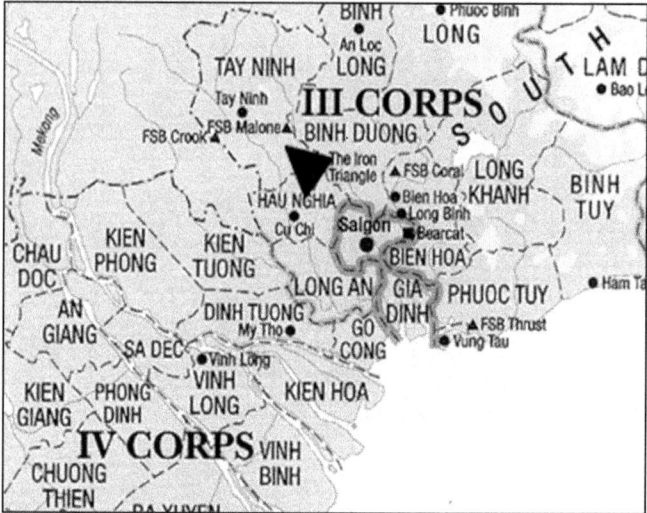

A map of Vietnam clearly showing the Iron Triangle just below and slightly left of the III Corps designation. There and in the surrounding area is the location of 7th Artillery operations from 1965 through 1970.

Early History in Vietnam of the 1st Battalion 7th Artillery the "Pheons"

Arriving on 23 June 1965, Private First Class Gerald Worster stepped off the plane at Saigon and became the first "Big Red One" soldier to set foot on Vietnamese soil. Worster was a member of the 1st Battalion, 7th Artillery's advance party. The battalion, along with Battery C, of the 8th Battalion, 6th Artillery attached, was commanded by Lieutenant Colonel John H. Frye III.

The 1st Battalion, 7th Artillery, along with the 2nd Brigade, departed Fort Riley for movement by rail to Oakland, California. On the 23rd of June the battalion loaded aboard the USNS Gordon and two days later sailed for Vietnam.

On 12 July, C Battery, 1st Battalion, 7th Artillery landed at Cam Ranh Bay, in the Republic of Vietnam as part of a taskforce with the 1st Battalion, 18th

Infantry Regiment, 1st Infantry Division. This force was the first tactical US Army unit to be deployed to the Republic of Vietnam directly from the continental United States. Their initial mission was to conduct limited offensive operations against the Viet Cong. On July 16 the battalion arrived in the vicinity of Bien Hoa and established a base camp.

During the weeks following arrival, the brigade area was the target of numerous Viet Cong probing attacks. The counter-mortar radar section, led by Chief Warrant Officer George R. Gurney, proved its worth as it rapidly located hostile mortars firing on elements of the 2nd Battalion of the 18th Infantry Regiment. Battalion FDC[1] promptly took action on the radar's data and the enemy mortars were silenced by counter fire. Throughout the latter part of July, August, and September, the batteries of the battalion operated in support of the 2nd Battalion 16th Infantry Regiment on battalion sized operations. On 26 September, C Battery departed Cam Ranh Bay with the 1st Battalion, 18th Infantry and rejoined the brigade at Bien Hoa on the 28th of September.

From September through October, the 1st Battalion, 7th Artillery detached the counter-mortar radar section and one firing battery to the division's staging area as security for the deploying 1st Division near Saigon University.

From 4 to 24 October the battalion, minus C Battery and the radar section supported the 2nd Brigade in a search and clear operation near Phuoc Vinh. This operation paved the way for the division's 1st Brigade, which established its base camp at Phuoc Vinh in late October. It was during this operation that the *Pheons* suffered their first killed in action, Private First Class James M. Long of B Battery.[2]

---

[1] Fire Direction Control.

[2] From Baltimore, Maryland, killed in action on October 5, 1965.

From 9 November to 21 November, the battalion, minus A Battery and the radar section participated in a clearing operation near Di An.[3] This action and a similar clearing action near Phu Loi led to the establishment of base camps for both the division headquarters and division artillery headquarters respectively.

Throughout this period, individual batteries operated in support of infantry battalion sized combat operations.

The heroic traditions of the 7th Artillery continued, as evidenced by the actions of Specialist Four Gunther Uhlman of Headquarters Battery. In August of 1965 Specialist Uhlman was working as a medic with B Battery during a search and clear operation. A Vietnamese soldier detonated a hand grenade booby trap and was severely wounded. Specialist Uhlman, himself wounded by the grenade, disregarded the presence of additional booby traps and rushed to the aid of the injured ARVN[4] soldier. He refused aid for himself until the Vietnamese soldier was properly treated and evacuated. For this act Specialist Uhlman was awarded the Soldier's Medal.

Vietnam Veterans Memorial wall # P-02E/L-108.
[3] Pronounced "Zeon".
[4] Army of the Republic of Vietnam.

The Soldier's Medal

The *Pheons* civic action efforts on behalf of the Vietnamese people were keynoted by Army physician, Captain Ronald P. Mahoney, the battalion surgeon. Doctor Mahoney spent long hours applying his professional knowledge to easing the pains and curing the numerous ailments of the Vietnamese peasants. Additionally, he made frequent flights to Ben San Leprosarium on the edge of War Zone D and established regular sick call for the lepers. Captain Mahoney's project for the battalion was building of latrine facilities to be installed at schools in the Bien Hoa area.

In the early part of December another member of the 7th Artillery distinguished himself. Sergeant Arthur Soto, was serving as an artillery reconnaissance sergeant[5] with Company C, 2nd Battalion, 18th Infantry during a search and destroy operation against the Viet Cong. At approximately 1630 hours on December 10th, the company encountered heavy fire from well concealed enemy bunkers and was forced to take cover. Sergeant Soto had reached safety with the last elements of the unit when he heard Sergeant Alford E. Scott Jr., who was wounded and could not move, calling for help. With total disregard for his own safety, Sergeant Soto immediately exposed himself and rushed to the aid of his wounded comrade and carried him on his back to safety. During this act Sergeant Soto was wounded by the continuing enemy fire. Even though wounded, Sergeant Soto insisted that Sergeant Scott be the first removed from the danger area. This outstanding display of aggressiveness and devotion to duty earned Sergeant Soto the Bronze Star Medal for Valor in ground combat.

---

[5] Often called a "forward observer".

Bronze Star Medal for Valor

From 16 December to 23 December the battalion, minus A Battery which was stationed at Di An, conducted a search and clear operation along Route 15 in the Long Thanh area. In this operation several Viet Cong base camps were found and destroyed with the help of fires from the 7th Artillery.

From 28 January 1966 to 15 February the battalion supported the 2nd Brigade again in an operation along Route 15 in the same area. The mission was to clear the area for the newly arrived 1st Brigade. This operation was code named MALLET.

The battalion then returned to its base camp in the Bien Hoa area to continue preparations for the move of its base camp to the new Bearcat area.

These preparations were delayed by the first full division sized operation code named MASTIFF, from 19 February to 27 February. This operation was the first time since its arrival that the entire battalion was airlifted. At the close of Operation MASTIFF, the air mobile portion of the battalion had a two day break at the base camp of the 2nd Brigade, 25th Infantry Division. Then on the 1st of March the air mobile portion of the battalion deployed from Cu Chi by CH-37 (Mojave) helicopter to an area south of Tay Ninh. Here the battalion had to be completely resupplied by air which was directed efficiently by Captain Charles

E. Martin, the battalion S-4 (quartermaster) officer. On the 5th of March the battalion was again moved by CH-47 (Chinook) helicopter to a Special Forces base camp outside Tay Ninh.

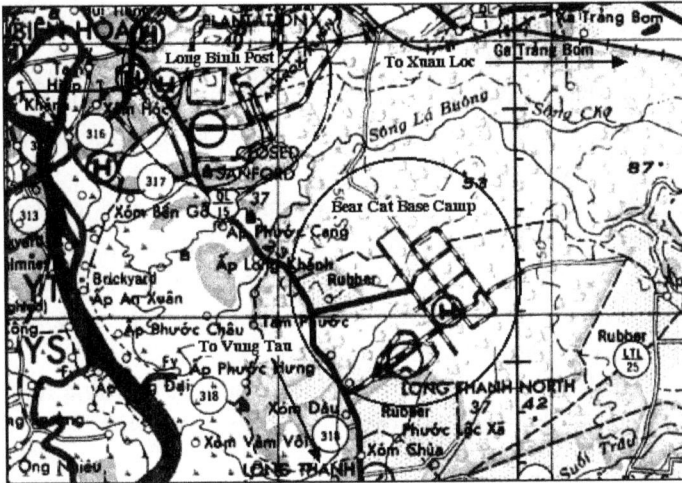

Map of the Bear Cat area, 1969

The move from Tay Ninh was by an Air Force C-123 and the command arrived at Bien Hoa Air Base that afternoon. Upon arrival final plans were made to move the entire battalion to its new base camp area at Bearcat. Everyone got in the swing of taking apart the buildings in the old area so they could be salvaged and reused in the construction of the new facility. While part of the battalion personnel remained at the Bien Hoa base camp, a portion of the unit moved to Bearcat.

Upon arrival the large task of clearing areas for concrete pads for the tents began. Work continued until the 29th of March when the battalion left for a staging area near Bien Hoa in preparation for Operation ABILENE. The next morning the battalion left in the direction of Vung Tau, and supported the brigade in that area until 14 April, 1966.

Crossing the Pond, December 1968

My war in Vietnam began in December 1968 and would last until January of 1970 with the entire year of 1969 sandwiched in between. I enlisted voluntarily in October of 1967 for an initial period of three years of active duty in the U.S. Army. Like all men at least aged eighteen, I owed a military obligation of six years to Uncle Sam. My first few months were taken up by basic combat training at Fort Polk, Louisiana also known as "Tiger Land" because it was the home of the Vietnam War infantryman.

I well remember the aromatic smell of the pine tree forests on that part of Fort Polk called "North

*Basic Combat Training October to December 1967
Fort Polk, Louisiana*

Rick Meyers, David Lemon, me, and Dave Edgar were all from Oklahoma. I grew up with Meyers and Lemon in Perry and we enlisted together. I would see David Lemon in Vietnam for the last time. He was murdered in 1975 after returning home to Perry, the crime is still unsolved. Both he and Rick were MPs in the Army. Rick joined the Oklahoma Highway Patrol after his military service.

Delta Five-Two

I was trained in D Company, 5th Battalion, 2nd Basic Combat Training Brigade. Marching was a big part of training at "Tiger Land". Note we are carrying the M-14 rifle that fired a .762 caliber NATO round, a much larger bullet than the .223 fired in the M-16.

Tiger Land 1967

Basic training was followed by three months of additional advanced training at Fort Sam Houston Texas where I learned the job of a medical corpsman. It was during this latter period of training that the dangers of my new profession became clear.

Between basic and advanced training I returned to my home town of Perry, Oklahoma and married my sweetheart, Carol Sellers. Originally from Texas, Carol was born in the small town of Temple just outside the gates of Fort Hood where I was born. When I reported to my Texas assignment I took Carol with me. Carol had family in Texas and we spent some pleasant weekends with them when I could get away from the post. But in January of 1968 the *USS Pueblo* was seized in international waters by vessels of the Democratic People's Republic of Korea. We were placed on alert which meant that training could be quickly curtailed and we would find ourselves on the way to a second Korean War.

The *Pueblo* is a Banner class technical research vessel under the command of Navy Intelligence and in fact is a spy ship. Today, the *Pueblo* is still held

Basic training dress uniform portrait 1967.

by North Korea although it remains a commissioned U.S. Navy ship.

On January 5, 1968, *Pueblo* left for Sasebo, Japan. On January 11 she departed Sasebo with orders to intercept signal traffic and conduct surveillance of Soviet naval activity in the area. In addition, she was to eavesdrop on the electronic signals of the North Koreans.

The crew was held for eleven months and slowly the tension was defrayed as other more pressing events in this turbulent year took center stage.

On January 30, the Viet Cong and North Vietnamese Army (NVA) launched an assault that became known as the "Tet Offensive". The liberal press, led by Walter Cronkite of CBS News, touted the offensive as the death knell of American efforts in Vietnam. In truth the offensive was a military defeat for the Communist North, but what they failed to achieve on the battlefield was won in the U.S. media and the minds of the American people. From that time on, Vietnam was seen by the majority of the people as a lost cause.

Meantime I finished my medical training and received my initial Army assignment. I was ordered to Fort Benning, Georgia and reported to

Training with the M-14 rifle at Fort Polk 1967.

the 690th Medical Company, an ambulance unit that provided medical support to training commands on the post, in March of 1968. While assigned with the company I performed medical "commitments" to the airborne, ranger, and Special Forces training commands as well as basic combat training units and the Officer Candidate School. Often I covered several firing ranges, sitting in my ambulance beside a telephone mounted on a pole. There I sat on call in case of a medical emergency. It was pretty easy duty with a lot of time to pursue my favorite hobby of reading.

Carol and I moved into our first home, a small trailer located in a park on the south side of the town of Columbus, Georgia. It was small, but it was home, and more importantly, it was what I could afford on the pay of a private first class. Meantime, other events were taking place in rapid succession.

On April 4, two weeks before my nineteenth birthday, Martin Luther King Jr. was assassinated. Once again I found myself on alert and I spent several days and nights in my ambulance awaiting a possible call to Washington D.C. to help in quelling the race riots ongoing there. The seriousness of the situation was underscored when we were issued shoulder holster Colt .45 automatic pistols. Except in the arena of politics the immediate situation calmed down.

Early one morning while in the motor pool cleaning my ambulance after an all night medical commitment in the field, the company commander, a captain, and the company first sergeant, paid me a visit.

I came to attention and reported to my captain, who directed me to carry on. While I worked on cleaning my vehicle both he and the first sergeant questioned me on basic military subjects, medical procedures, current events, and knowledge of my ambulance and its maintenance. I didn't realize at the time that they were in fact holding a promotion board.

Both seemed to be pleased with my answers and a week later I was promoted to the rank of specialist E4 with a nice pay raise. Carol and I moved into a larger trailer. However, I still could not afford a car, and I continued hitchhiking from the post to home after each duty day. Many soldiers driving back and forth to the fort were Vietnam veterans. Upon seeing the red cross brassard on the arm of my fatigue uniform they would invariably stop and give me a ride, referring to me as "Doc" in our conversations. It was a nickname I would grow fond of and grew used to during my tour of duty in Vietnam.

In June of the year the nuclear submarine *USS Scorpion* was lost off the Azores while observing Warsaw Pac naval maneuvers. Although kept from the American public, the U.S. Navy suspected the Russians sank the *Scorpion* and quietly went on alert, turning up the burner on the Cold War. Many years later some historians and Navy officers believed the *Scorpion* was sunk with the help of Navy warrant officer and spy John Anthony Walker Jr.

The Soviet action was believed to be in retaliation for the sinking of their nuclear missile submarine the *K-129*, a few hundred miles from the U.S. naval base at Pearl Harbor, Hawaii. The *K-129* was many hundreds of miles from her assigned area of operations at the time of her loss. Many years later in the non-fiction book "Red Star Rogue", authors Kenneth Sewell and Clint Richmond advocate that *K-129* was a rogue ship that left her assigned area with the intent of perpetrating a nuclear attack on Hawaii. They go on to claim that the U.S. Navy discovered her sinister mission and intentionally sank the *K-129*.

Shortly after the loss of the *K-129* the Soviets observed the *USS Swordfish*, another American submarine, arriving in Japan with a damaged sail and periscope. The Soviets began to suspect their submarine was destroyed in a collision with the *Swordfish*. All of this added up to a growing military and political tension between the U.S. and the U.S.S.R.

Despite the American military victory during Tet, more soldiers were ordered to Vietnam, the 690th received an alert that several of their medics would be shipping out in the next few months. I had no idea I would be slated to go along.

In August of 1968 the Russians invaded Czechoslovakia and the 690th went on alert again. A street slogan in the capitol of Czechoslovakia read: *"Lenin wake up, Brezhnev has gone mad."* A reference to the then Soviet Premier, Leonid Ilyich Brezhnev who ordered Russian and Warsaw Pact troops to invade and remove the liberal Czech leader Alexander Dubček and his reformist administration.

Once again the Cold War was turned up a notch, but it had no bearing on what was happening to me and several other trained medics at the 690th Medical Company. In November, I was notified of my pending reassignment to Vietnam. I spent most of that month going through RVN training.[6]

We spent a few days conducting combat training operations in a mock-up Vietnamese village. There we learned about booby traps, weapons caches, and enemy hiding positions called "spider holes". We also took "quick kill" training to learn to fire from the hip by instinct rather than by aiming. Initially, we began with bb rifles, hitting large metal slugs and eventually dime sized objects tossed in the air. Later we graduated to hip firing the M16 rifle. I was very good at it. We went through night time operations and were exposed to the eerie dangling red glare of parachute flares. There was additional instruction in the use of Claymore mines and trip flares attached to barbed concertina wire and classes on camouflage. Trained and motivated, I was ready to go, but not looking forward to a yearlong separation from Carol.

I received my orders toward the end of the month and Carol and I said goodbye to our first home and made our way to Texas. We decided that she would

---

[6] Republic of Vietnam.

live with her folks in Houston while I served my tour of duty. Carol and I, seeking a little time alone together for what might be the last time, spent some time in Galveston, walking its beaches and talking about the future. We both knew there might not be any future for the two of us. During my time in the war I often thought back on our time spent in Galveston. This was especially true when Glen Campbell released the song "Galveston" about midway through my combat tour. His lyrics seemed to have been written just for us. Carol was twenty-one, when I left Galveston. It remains one of our all-time favorite songs.

Galveston, oh Galveston, I still hear your sea winds blowin'
I still see her dark eyes glowin', She was 21 when I left Galveston

Galveston, oh Galveston, I still hear your sea waves crashing
While I watch the cannons flashing, I clean my gun and dream of Galveston

Galveston, oh Galveston, I am so afraid of dying
Before I dry the tears she's crying
Before I watch your sea birds flying in the sun
At Galveston, at Galveston...

Before leaving for Oakland Army Airbase in California, I spent a few days with my parents and brother and sisters in Ripley, Oklahoma and with my older brother Don in Tucson, Arizona.

The day after Christmas, in that turbulent year of 1968, I climbed aboard an airliner in Houston and with Carol and her family waving goodbye from the terminal; I began my journey to America's longest Asian war.

"Carry on my wayward son...there'll be peace when you are done. Lay your weary head to rest. Don't you cry no more."

-Kansas

SP4[7] James Cribbs and Walt Cross at the replacement depot, Vietnam. Cribbs and I served together in the 690th Medical Company before coming to Vietnam together. He was assigned to an infantry unit.

Vietnam

I arrived in the Republic of Vietnam at the end of the turbulent year of 1968. The Vietnam War is also known as the Second Indochina War, the Vietnam Conflict, and among Vietnam Communists, the American War. It officially occurred from March 1959 through the end of April 1975. The war was a struggle between communist dominated North Vietnam with its capital in Hanoi, and the U.S. supported Republic of Vietnam in the south with its capital in Saigon.[8]

Prior to World War II Vietnam was a French colony and known as French Indo-China. During the conflict of World War II French Indo-China fell to the

---

[7] Specialist Grade Four.
[8] Later renamed Ho Chi Minh City by the communists.

surging military forces of Imperial Japan. This demonstrated inability of France to defend her colony was noted by political forces that wanted to throw off the rule of the French. At war's end these political forces coalesced into the communist dominated Viet Minh. The forces were set for the first Indo-China War that ended in the French defeat at Dien Bien Phu.

The French met with the Viet Minh in Geneva, Switzerland and an accord was reached where the communists would restrict themselves to the north while the French would hold the south until elections could be held.

The U.S., sensing the communists would win the elections, began to groom Ngo Dinh Diem, a Catholic, as the next leader of South Vietnam. A man of courage and a nationalist, Diem spent many years in the U.S. and was acquainted with Senator John F. Kennedy. But when Kennedy became president he saw that Diem needed to be replaced. His generals staged a coup and Diem was assassinated. Kennedy soon shared the same fate and U.S. involvement in Vietnam grew deeper.

NUNQUAM FRACTUM

The 7th Artillery's Distinctive Unit Insignia. Nunquam Fractum is Latin for "Never Broken".

On the 23rd of June 1965, PFC Gerold Worster arrived and stepped off the plane near Saigon, the capitol city of the Republic of Vietnam. Not only was Worster the first soldier of the "Big Red One" to set foot in Vietnam, he was also the first soldier of the 7th Artillery. At that time the 1st Battalion, 7th Artillery was commanded by Lieutenant Colonel John H. Frye III. During my tour of duty the battalion would be commanded by three other Lieutenant Colonels; Francis King, Charles C. Sperow, and Raul A. Garibay. Garibay would later retire a full colonel and make his residence in El Paso, Texas. Sperow would also reach the rank of full colonel and after retirement make his home in Florida.

The 1/7th Artillery, along with the 2nd Brigade of the 1st Infantry Division departed Fort Riley for movement by rail to California. The same day PFC Worster arrived in Vietnam, the battalion loaded aboard the USNS Gordon and forty-eight hours later sailed for Vietnam.

I arrived in Vietnam near the end of December, 1968 and spent New Year's Eve in the enlisted and NCO[9] club at the replacement depot not far from Tan Son Nhut airbase, where I had arrived. It mattered not that I was only nineteen. I drank a few beers and listened to music played on the radio by the American Forces Vietnam Network. A hand painted sign behind the bar paraphrased General Tecumseh Sherman's famous statement of the Civil War in a way never envisioned by him, it proclaimed "War is hell, but combat is a motherfucker". I was soon to find out what that meant.

---

[9] Non-Commissioned Officer.

I reported to Headquarters and Headquarters Service Battery, 1st Battalion, 7th Artillery of the 1st Infantry Division located in Di An a few kilometers north and east of Saigon.

I was greeted by a gray haired and to me, a very old staff sergeant. Quite likely he was only in his mid to late forties. Speaking with him that night he told me he had come out of retirement and volunteered for duty in Vietnam with the hopes of obtaining a promotion to sergeant first class. He lamented however that his tour was nearly up and it appeared the promotion he was hoping for was not in the offing.

He showed me where to stow my dress khaki uniforms, I wouldn't need them in the field, and I donned my new jungle fatigues. These were olive drab in color, the same as the duty fatigues I wore stateside, but that was where the resemblance ended. Name tags, sleeve rank, and the "U.S. Army" above my right shirt pocket were all in black letters. The shirt was not tucked into the trousers, but worn loose and had large pockets in the front and sides for storage. The trousers too had two large cargo pockets located on the thighs. Curiously, the 1st Infantry Division patch worn on the left shoulder, displayed its original colors of olive drab with the number "1" centered and colored red. It is from this distinctive numeral that the division takes its nickname "The Big Red One". The olive drab of the patch was of a different shade than the uniform and so it stood out. The 7th Artillery pocket patch, worn on the top left pocket of the uniform near the heart, was in full color of silver and red. I recall thinking it made a very good target centered over my heart.

"Get a good night's rest son, in the morning you will report to the brigade jungle school." The old sergeant told me. When I inquired what the school was about he explained I would learn about booby traps, trip wires, and other favorite tricks of "Charlie".[10]

I went to bed in the headquarters building that night, not particularly looking forward to more training. I shouldn't have worried about it. About 0300 hours[11], in the morning, the staff sergeant woke me up. "Get up Doc." He said, "They want you in the field ASAP. Battery A was hit last night, and the medic was wounded. You are his replacement anyway and they're sending him in from the field." He continued. "Guess he did a good job, got a Army Commendation Medal with a "V" device."[12]

He hustled me off to breakfast and then down to the supply sergeant who was wiping the sleep from his eyes. He handed me a helmet and a blood- stained flak jacket. Unlike the vests worn today in Iraq or Afghanistan, the flak jackets of 1968 were not bullet proof. They offered the minimum protection from shrapnel, but did nothing to slow down a bullet. The supply sergeant saw me looking at the stain and took a drag off his cigarette. "Yeah, he didn't keep it zipped up. A chunk of mortar round got him. You make sure you keep it zipped Doc." He said, then he leaned over and pulled up an ancient looking Thompson submachine gun. "You want this Tommy gun Doc?

---

[10] A nickname for the Viet Cong derived from the phonetic alphabet used for clarity on military radios. "Victor Charlie" was a synonym for the acronym VC, which in turn meant Viet Cong. We sometimes referred to them in a singular term as "Mr. Charles". The Viet Cong were the Communist-led forces fighting the South Vietnamese government. The political wing was known as the National Liberation Front, and the military was called the People's Liberation Armed Forces. Both the NLF and the PLAF were directed by the People's Revolutionary Party (PRP), the southern branch of the Vietnamese Communist Party, which received direction from Hanoi.

[11] 3:00am

[12] In the Army, a "V" device indicates the award was given for personal valor in combat.

Could be pretty good at short range." He said in an earnest voice.

I had no doubt my dad, a World War II veteran, would have known how to use it, but I didn't have a clue. I told him I'd better stick with the M16 I knew how to use. He sighed and put it back under the counter. "Nobody wants the classic stuff anymore." He groused as he handed me a rifle along with four ammo clips. I also took a bayonet and an M1911A1 .45 caliber Colt automatic pistol and its black leather holster stamped with a large "U.S." on its flap. I would wear that pistol for more than a year and when I finally went home I would feel naked for months afterward without its comforting weight hanging on my right side.

The holster had a leather tie down I used to secure it to my leg like an Old West gunfighter, and that kept it from rising up when I drew the pistol. I had pause to wonder why I had been trained on so many weapons before I shipped out to Vietnam except the one I would use the most, the .45 caliber Colt. Only the medic, the battery commander and the first sergeant wore pistols.

Sparky Fence Alpha[13]

After a final cup of coffee and a stop at the ammunition point to draw rounds for my rifle and pistol, I boarded a Huey[14] helicopter for the first time and flew to my new unit, Alpha Battery, 1st Battalion, 7th Artillery currently in the division's forward headquarters at Lai Khe.

A few days after I arrived the battery displaced and moved on the 16th of January to FSB Huertgen near

---

[13] This was the code name for Alpha Battery, 1st Battalion, 7th Artillery.

[14] Also known as a "slick".

the Iron Triangle in support of the 2nd Battalion, 16th Infantry. A short time later we moved on to FSB Aachen where we would remain until the end of April. It was at Aachen that I celebrated my birthday on April the 20th and at the ripe age of twenty I left my teenage years behind. It was in April that the troop strength in Vietnam would peak at 543,400.

When I got "in country" the latest campaign, designated *Counteroffensive Phase VI* dated from November 2, 1968 to February 22, 1969 was in full gear. During this time period there was a marked increase in enemy activity in the area known as the Saigon Rocket Belt. This gave an indication that the enemy was likely preparing to launch an offensive comparable in scale and magnitude to the 1968 Tet[15] Offensive.

In anticipation of such an attack, elements of the 2nd Brigade, 1st Infantry Division, increased their operations in this area. Supported by three firing batteries each composed of six 105mm howitzers of the 1st Battalion, 7th Artillery, the maneuver battalions of the 2nd Brigade were able to initiate a counteroffensive. The actions of the brigade thwarted the enemy forces of Viet Cong and the North Vietnamese Army (NVA). Ralliers[16] continually supplied information from which enemy targets were identified and fired upon, an effective part of the 7th Artillery's intelligence and interdiction program.

During the course of these operations the 7th Artillery's firing batteries displaced and moved between base camps and fire support bases more than 40 times during the year of 1969. These fire support

---

[15] Short for Tet Nguyen Dan, it is the most important Vietnamese holiday. Tet is the celebration of the beginning of spring as well as the New Year. It is the time of family reunions and exchanging gifts.

[16] Viet Cong and NVA defectors.

bases, with names from states and of famous 1st Infantry Division battlegrounds and unforgettable division personalities, live on in the memories of the men of the 7th Artillery. From these bases the firing batteries of the battalion supported the maneuver elements of the 2nd Brigade.[17]

On January 1 Battery A was located in the 1st Division's forward base in Lai Khe[18] while B Battery was positioned at FSB Holiday Inn; and C Battery was located in the village of Cat Lai. Two heavy mortar platoons were attached to the battalion and located at Di An and Phu Loi.

When I arrived Alpha Battery was under the operational control of the 2nd Battalion, 33rd Artillery; Bravo was supporting the 1st Battalion, 18th Infantry of the division and Charlie Battery was west of Saigon in the Rocket Belt.

The two heavy mortar platoons were providing close-in fire support for their respective base camps. During the year one platoon would remain in Di An while the other rotated between Phu Loi and FSB Thunder II.

February had seen the end of my first campaign in Vietnam and the start of the *Tet 69 Counteroffensive* that ran from February 23 to the 8th of June and was the 7th campaign of the war. Unknown to me, I had already earned two campaign stars on my Vietnam Service Medal. I would earn two more before my tour ended.

Unknown to us, President Richard Nixon, in an attempt to cut the Ho Chi Minh Trail, had ordered the secret bombing of Cambodia in March of 1969.

---

[17] The maneuver battalions of the 2nd Brigade consisted of 1st and 2nd battalions of the 16th Infantry Regiment and the 1st and 2nd battalions of the 18th Infantry Regiment.
[18] Enroute to FSB Huertgen.

Also in March, Bravo Battery split and sent three guns to our future home near Phuoc Vinh, Fire Support Base Jim. Bravo's other three guns proceeded on to FSB Riley but soon moved on again to FSB Yakima at the Thu Duc water plant.

My battery was situated on the southeastern corner of the Iron Triangle. Located a short thirty kilometers northwest of Saigon, it was to the west of the village of Ben Cat and dominated a section of Highway 13.[19] This important national highway stretched to the Northeast from Saigon to the Central Highlands, ending at the regional center town of Ban Me Thuot. Due to the nearly constant combat along this highway it was known to American soldiers as "Thunder Road". And the fire support bases along this road were called the "Thunder" positions.

This dangerous and mysterious area was and had always been a VC strongpoint. The fifty square kilometer area was famous for its stern resistance to American incursions and the ARVN[20] avoided it whenever possible. The Iron Triangle was riddled with Viet Cong tunnels and underground bunkers that provided the enemy space for medical aid stations, supply depots, training rooms, and unit headquarters.

The first American attempt to subdue the area was carried out by the 173rd Airborne Brigade almost four years before in October of 1965. The brigade swept the triangle near where the 1st Infantry Division was to make their home near the village of Lai Khe.

---

[19] FSB Aachen was approximately 13 km west southwest from division headquarters at Lai Khe, 2 km north west of the village of Xa Bung Cong, and 8 km due west from the village of Ben Cat. It was named after the 1st Infantry Division WWII battle site.
[20] Army of the Republic of Vietnam.

It was literally an impossible task. Following a heavy artillery preparation and air strikes, the 173rd air-assaulted by helicopter into the Iron Triangle. The brigade was joined by the 1st Royal Australian Regiment and they struck the north side of the triangle and pushed south, driving the Viet Cong before them, or so they thought. Although they found some caches of weapons and supplies and made contact with the VC, most of the enemy soldiers simply went underground. The Iron Triangle would remain a Viet Cong stronghold and in 1969 I had firsthand experience of it.

Movin' Into the Wrong Neighborhood

Early in March we were told the 16th Infantry Regiment (1st Infantry Division) was moving into position a few klicks (kilometers) north of our position. They were to establish a new base of operations, so they were accompanied by combat engineers. They spent the entire day bulldozing a berm or earthen work around which they cleared away the jungle and began laying concertina (barbed) wire. As they laid the wire, they set up trip flares, a type of warning device, in between the strands of wire. These flares would go off at night if someone attempted to spread the strands apart. They also planted claymore mines, careful to turn the side that read "this side toward enemy" out toward the jungle. Some of the men gathered "C" ration cans and attached them to the wire with small stones inside them that would hopefully make noise if Charlie messed with the wire.

We were on standby in case they needed help as they worked feverishly to complete the position and close the wire before the onset of sudden nightfall.

They almost made it, but it wouldn't have mattered if they had, for it was later determined that they had built their new base atop a VC regimental headquarters. As the shadows of the rubber trees stretched out to grip them in their dark talons, the enemy rose from the ground like phantoms. The fighting was hand to hand, grenade to grenade, rifle butt to rifle butt. The sound of small arms fire came to us, the quick zipper sound of M-16s, the deeper stutter of AK-47s and the deadly rattle of M-60 machine guns.

The "thunk" of enemy mortars and the similar sound of M-79 grenade launchers could be heard, followed by the crash of the exploding round. Calls for help came over the radio waves as we stood by helpless, for the enemy was among our own. The fighting continued for what seemed like hours as we fired illumination rounds to bring light to the thick darkness where men struggled and died. Slowly, slowly, the 16th disengaged and while one gun continued to fire illumination, the other five cranked up high explosive rounds and finally we provided cover fire for the retreating infantry and their supporting engineers. The next morning FSB Aachen was used as a collecting point for our dead. I got my first glimpse of body bags being unloaded off vehicles and loaded onto slicks for transport to the mortuary unit. It was a grim sight.

We continued firing most of the night, while spooky (C-47 gunship outfitted with a mini-gun) circled above. But the enemy had done their work, and slipped silently back underground. By the time a major offensive could be launched two days later, the VC regiment had slipped away, melting into the jungle. The Iron Triangle remained firmly in communist hands.

Checking Ammo

The battery commander at the time of my arrival was Captain John Dubia[21], he and the executive officer, a 1st lieutenant, were both West Point graduates. They were extremely proud of that fact, as indeed they should have been.

I recall an incident involving the XO that was both amusing and showed how versatile the lieutenant was. Late one night the battery received a fire mission to support an infantry ambush that was ambushed itself. It was extremely dark and the only light came from red lensed flashlights. The monsoons were in full swing, and rain was pouring down. The lieutenant was strutting around on top of the sandbag parapet that surrounded and protected each gun and provided a storage place for ammunition.

Coordinates came down the land line from the fire direction control center (FDC) and were repeated to the gun crew. The "Chief of Smoke", a sergeant first class, was supervising the mission.

The guns crashed as the command to fire rang out. At that moment, with rain slicked sandbags and the concussion from the roar of six 105mm howitzers going off in unison, the first lieutenant slipped. He did a perfect 180 degree flip and landed inside the gun position directly on his butt, his helmet plopping upside down in the mud beside him. There was the sound of muffled chuckling from the gun crew and then the chief asked the West Pointer what he was doing. Recovering quickly, the lieutenant shined his flashlight on the wooden ammo boxes beneath the parapet and replied in a cool voice. "I'm checking the

---

[21] John Dubia would rise to the rank of a full (4 star) general. Four officers I served with during my 21 year career advanced to the rank of general officer. At the time we served together they were captains or majors.

ammo, chief." His quick thinking retort earned him grudging admiration and there was no more laughter.

Most people are familiar with the French tire manufacturer, Michelin. They are probably less aware that Michelin owned a large rubber tree plantation in Vietnam. For that reason, every rubber tree we damaged with our artillery while fighting the communists, Uncle Sam paid the French.

One day in April we got into a hot fire mission as NVA (North Vietnamese Army) and Viet Cong units brought the large base at Lai Khe[22] (pronounced Lie Kay) and the location of our division's headquarters, under assault. Our battalion commander was aloft in his Loch[23] and was directing suppressing fire[24] from the air.

The colonel gave Alpha Battery a "Fire at will" order and we took him seriously. I was humping ammunition to the guns right alongside the artillerymen. The battery was firing as fast as it could get the coordinates laid in. The mound of brass shell casings beside each gun continued to grow apace as shell after shell was rammed home, the breech closed, and the order to "fire" called out by the gun chiefs. Rivers of sweat poured from us as we answered the call for help. Somewhere I heard an artilleryman cry out "You yell, we shell, like hell!"

I was so busy carrying shells that I failed to notice the gun behind me traverse.[25] Unknown to me, the howitzer's muzzle was just over my head. As the gun fired, the concussion blew my helmet off my head,

---

[22] Lai Khe is located 40 km due north from Saigon and 22 km southwest of Phuoc Vinh where David Lemon was stationed. It was the home of the 1st U.S. Infantry Division and the 5th ARVN Infantry Division as well as the 161 Battery, Royal New Zealand Artillery.

[23] Reconnaissance helicopter.

[24] Counter mortar, counter rocket fire.

[25] Change direction of aim.

and I staggered beneath the weight of the two 34 pound rounds on my shoulders and the effects of the explosion. I remained erect and delivered the rounds to the gun, but my ears rang for three days afterward and caused me some loss of hearing later in life.

Finally, after half an hour of six guns firing as fast as possible, the order came to cease fire. As we stood trying to catch our breath, a chopper came down and settled in our fire support base, kicking up a large cloud of dust. We cursed the idiot causing this inconvenience and then watched as the 7th Artillery Battalion Commander, Lieutenant Colonel Charles Sperow, climbed out.

Our captain seemed excited as he greeted the colonel and they shook hands. I later learned that after the mission the colonel had radioed each battery for a "rounds expended" report. The other two batteries engaged besides us had fired a little over 200 rounds each. Alpha Battery had fired over 700 rounds, depleting our basic ammunition load. The colonel had to order us an emergency re-supply by chopper.

After admonishing us for shooting up our combat load, the colonel grinned and said "You guys have got to be the fastest gun in the Big Red One. But I'll never give you a fire at will order again." Alpha battery felt justifiably proud, we had responded to the needs of our comrades, and had done it in style.

High Flyin' Cowboy

"High Flyin' Cowboy this is Sparky Fence Alpha, over." The radio operator in the "exec post" released the key to his mike, awaiting a reply.[26]

---

[26] I have used a literary device here to paraphrase the actual communication. Normally the battery did not have direct radio contact with aircraft.

"Sparky Fence Alpha, High Flyin' Cowboy, what d'ya redlegs need down there?" The gunship commander replied.

"High Flyin' Cowboy, we are receiving enemy mike[27] fire, location unknown, and azimuth[28] changing, over".

"Roger Sparky Fence Alpha, we're on the way." And the modern day cavalry, troopers in high flying steeds of olive drab sporting the crossed cavalry sabers painted yellow upon the nose of their ship, sped to the support of Alpha Battery. For several days now, American positions along "Thunder Road" were coming under VC mortar fire, a fire that moved quickly and struck again and again from different positions.

Thus far the enemy soldiers managed to avoid return artillery fire. Now, in the darkness above the tree-line, like winged predators searching for prey, came the air cavalry. Aerial spotlights stabbed through the darkness searching, searching, and finding. The three gunships making up the hunting pack began a Ferris wheel like rotation, one ship following another in a wheel of death as they came to the bottom of the circle, releasing a rain of red tracer marked death, followed by the roar and flash of rockets slashing the night sky.

"Sparky Fence Alpha, we have a target, engaging." The lead ship commander stated. Too late, a slender stitched line of green Chicom[29] tracers arced upward in a weak and futile attempt to engage the aerial sharks. The audacious but weak fire was swiftly answered as the mini-guns of the choppers engaged the enemy target with ship after ship sending a lethal rain of lead earthward. Each burst appeared to be a

---

[27] Mortar.
[28] The angle between a reference plane and a point.
[29] Chinese communist.

continuous line of red tracers, but in truth five rounds bridged the gap between each tracer and its twin further behind.

Now the ships paused, hovering, like giant eagles pausing and circling its prey. The searchlights played around the wood-line, a shower of white light in the blackness of the rubber trees. "Sparky Fence Alpha, it was a Lambretta[30] outfitted with VC troops and at least one mortar. They are neutralized." He continued, almost as an afterthought.

"Roger High Flyin' Cowboy, copy one VC vehicle with mortar and enemy casualties. Nice shootin' and thanks." The radio operator replied. And as the choppers reformed and flew over Fire Support Base Aachen, the redlegs cheered.

Medic!

Despite the fact I spent a lot of my time performing as an artilleryman, I was, never the less, the battery's medic. One night a blood curdling scream, followed by cries of "Medic, medic!" had me scrambling through the gun positions, aid bag in hand. There on the ground a soldier writhed in agony, both hands pressed to the right side of his head. As I arrived he screamed again, kicking wildly around. I yelled at two guys to help me hold him still and pulled his hands away so I could see his head. I had expected to see a chunk of metal in his skull, perhaps a flying fragment from the friendly shells falling around our perimeter. These shells were known as H&I, or harassment and interdiction. Their purpose was to catch the VC by surprise if they were sneaking up on our position. Or I thought perhaps he'd been headshot

---

[30] Three wheeled Vietnamese vehicle named for its Italian manufacturer.

by a VC sniper. But there was no sign of trauma or blood in fact I couldn't see a damned thing.

"What the hell's wrong with you?" I demanded as he screamed again. "In my ear, buzzing!" He cried. I looked deeper into his ear with my red lensed flashlight and could just make out the ass end of a small beetle. As I watched, he tried to fly deeper into the soldier's ear, his wings beating a furious tattoo inside the ear canal. I'd seen enough, and as the victim let forth another scream I poured hydrogen peroxide from my aid bag into his ear.

The bubbling solution took no time at all in sending the beetle floating up and out of the ear. Just then the captain arrived, a little breathless, along with the first sergeant. When I explained the situation, they both left shaking their heads and chuckling. The patient was very appreciative of my medical efforts.

Another time I was called to an emergency and found two soldiers lying on the floor of their tent in convulsions. They were moaning and thrashing about, and foaming at the mouth. I was perplexed, I could see one of them possibly having got past army doctors with a case of epilepsy, but not two, and they certainly wouldn't have had fits at the same time. I suspected poisoning of some kind.

I got the two men restrained on litters and insured they wouldn't swallow their tongues. We quickly got them to the evacuation point and medivac choppers took them away.

I searched their tent after they had gone to see if I could find the cause. I found a jiffy pop popcorn package, the kind that came already sealed in its own aluminum covered popper. One of the two had gotten it in a "care" package from home. It was mostly empty. I couldn't understand how they could have been poisoned by popcorn. Then I saw they had been heating the popcorn by burning C-4.

C-4 is a plastic explosive and is very effective at blowing things up. But when lighted with a match it burns like a can of Sterno. The problem was that the bottom of the lightweight aluminum popcorn popper had a pin sized hole in it. The fumes of the C-4 had gathered inside the closed popper and permeated the popcorn. When the two soldiers ate it, it was as if they had eaten C-4.

Both of them were hospitalized, one never to return. The second guy came back, but he was never the same again. I believe the C-4 caused both of them brain damage, one so bad he never came back. I had radioed my suspicions to Divarty HQ (division artillery headquarters) and they passed them on to the docs at the hospital.

Eight days later on the 28th of April, we loaded into trucks, hooked up the howitzers and displaced to another fire support base, this one with the simple name of Fire Support Base Jim.

Well to the north of us in the month of May, 1969 the "Battle for the A Shau Valley" was beginning and the phrase "Hamburger Hill" would reverberate in the United States.

We were moved to FSB Jim to support the Song Be Road Operation. This operation was to open and secure the road between Phuoc Vinh and Song Be to provide a main overland supply route between the two cities. FSB Jim sat south of this road, and our six guns were to defend the convoys moving up and down the road to Phuoc Vinh. Song Be was the logistical support base for the division's units operating out of the northern town of An Loc. To provide additional protection, three guns from Charlie Battery moved from Cat Lai to Di An on 8 May and then on to the village of Quan Loi. This was very far north of our position and about 90 kilometers from Saigon.

FSB Jim was located 12 kilometers east of the division headquarters at Lai Khe and two km south of the village of Phu Giao in the Binh Duong Province portion of III Corps' Zone of operation. The 1st Cavalry Division's field headquarters was just 13 kilometers north of Jim at Phuoc Vinh. There was also an airfield there with the original name of Phuoc Vinh Air Field. Phuoc Vinh was once a major base of the 1st Infantry Division until November 1968 when the 1st Cavalry Division assumed possession.

I visited Phuoc Vinh, it must have been around June or July of 1969 to pick up supplies and our vehicle was stopped by an MP at the gate. I sat in the passenger side and heard someone from the lookout tower yell out "Walt, Walt." But I had been called "Doc" for so long that I ignored it, thinking it was someone else named Walt. Then the voice yelled "Hey, Cross!" and I knew he was talking to me.

"He" turned out to be David Lemon, my old friend from Perry, Oklahoma. I had enlisted with David and Rick Meyers, all of us from Perry, and took basic combat training with them.

The year 1969 saw the largest buildup of American troops during the war, reaching a swell of more than five hundred thousand men. What are the chances that two men from the same small town in Oklahoma with a population of around 6,000 souls, would come across one another? I put the odds at hundreds of thousands to one if not more. Perhaps because of the fate that awaited David, Providence had a hand in our meeting.

David had another MP take his position in the tower and he and I had a good long visit and shared a few beers at the PX. I had no idea it was to be the last time I ever saw him, for he was killed a short time later. No, he didn't die in Vietnam, but in his bed as he slept in our mutual home town of Perry, Oklahoma after returning unscathed from the war. The crime has never been solved. Was it someone with a grudge at home, or did something sinister and deadly follow David home from Vietnam?[31]

Oddly enough, in November of 2008 I got a chance to look a little closer at events surrounding David's death. In the first week of that month I received a "Patriotic Employer Award" for Mr. Jim Davis, the city manager of my old home town of Perry.

These ESGR[32] awards are given at the request of a Guard or Reserve member in recognition of their employer's support for their military service. The nominator in this case was Master Sergeant Mike Thomas of the U.S. Air Force Reserve. Mike is a veteran of the Gulf War. Mike also happens to be the Chief of Police for Perry.

I called Mike and made arrangements to make the presentation to Jim Davis. We surprised him at city hall and the presentation was made. During our conversation I learned that Jim was a Vietnam veteran, serving there in 1968. I told him about enlisting with both David Lemon and Rick Meyers in October of 1967. I then told both Jim and Mike about the murder of David in 1975 and explained I would like to know more about the events surrounding the death of my friend. Mike told me he would look into it and let me know what he could discover. However, at the time of this book's publication he had not gotten back to me and so the mystery continues.

---

[31] There is at least one precedent where it is believed that an American soldier, in this case Colonel Nick Rowe, was pursued and assassinated outside Vietnam by communist agents for his personal actions during the Vietnam War. This assassination took place in April of 1989 in the Philippines. Could something similar have overtaken former military policeman David Lemon?
[32] Employer Support of the Guard and Reserve a program of the Department of Defense.

Beginning on May 25 Alpha Battery alternated three of our guns between FSB Jim and FSB Thunder II to support the 2nd Battalion of the 18th Infantry and simultaneously support Operation Thunder. This operation was initiated to provide security for convoys moving along Highway 13. On the 28th of June the battery reassembled at FSB Jim and shortly thereafter Captain Robert S. Mager was replaced by Captain Harry G. Madden as our battery commander.

About this time Alpha Battery's guns were replaced with a new howitzer, the M-102. It was lighter in weight than our old "split trail" guns and at first the men scoffed at it, referring to it as "Made by Mattel". Its characteristics are:

- Caliber: 105 mm (4.13 in)

- Length: 17.1 feet (5.2 m)

- Width: 6.4 feet (2 m)

- Height: 5.2 feet (1.6 m)

- Weight: 1.6 tons (4.7 t)

- Crew: 8 (about half that number in Vietnam)

- Rate of fire: 10 rounds per minute maximum, 3 rounds per minute sustained.

- Range: 11,500 m (7.1 miles), 15,100 m (9.4 miles) with rocket-assisted projectile (rocket assist was not available in Vietnam to the 1/7th).

It fired a 34 pound high explosive projectile with a variable fuse (air burst, ground burst, penetration burst) and was boosted by powder charges from 1 to 7 in strength depending upon the target's range.

The M-102 replaced the M-101 pictured below. Note the large split trails that steadied the older model gun.

In June of 1969 President Nixon ordered the first American combat troops out of Vietnam and back to the United States.

An M-101 in action in 1969.

This artilleryman screws a timing fuse into the nose of a 34 pound 105mm shell, June 1969. Photo courtesy National Archives.

July saw Charlie Battery move to FSB Venable Heights to provide closer support to the 2/18th Infantry. In August Bravo Battery, which was occupying FSB Cheyenne and Yakima with three guns each, reassembled at FSB Mortain. Despite research I have not been able to discover the location of these three fire support bases. However, I suspect that FSB Mortain was located north of Lai Khe and near the Thunder positions.

Only at FSB Venable Heights for a month, Charlie Battery now moved north to FSB Thunder II. There it came under heavy assault by NVA troops on September 5 during what became known as the "Battle of Thunder Road". This was an opening gambit that would lead to an enemy attack on my unit at Fire Support Base Jim.

In the early morning hours of September 5 the small village of Chon Thanh, located south of FSB Thunder III and defended by ARVN troops, came under attack. The village was subjected to an intense mortar attack followed by a considerable force of enemy soldiers operating as Sappers. Charlie Battery responded by allocating two guns to provide illumination fire. The enemy then attacked FSB Thunder III with the intent of stopping Charlie Battery's illumination mission. Mortar rounds landed in the base followed by Sappers probing the perimeter wire. A third gun was dedicated to providing illumination, this time for Charlie Battery itself, while still continuing the mission for the ARVN troops. The other three guns were directed at perimeter defense and began to fire point blank high explosive rounds followed by the deadly flechettes rounds called "Killer Junior". During this fierce exchange 24 enemy soldiers were killed, wounded, or captured. During this opening engagement Charlie Battery expended more than five hundred rounds of 105mm ammunition including illumination rounds.

The following morning a convoy led by the 2nd Battalion, 34th Armored of the 25th Infantry Division was attacked and Company C, 2nd Battalion, 2nd Infantry (Mech)[33] on FSB Thunder III, responded. 2d/2d was joined by 1st Battalion, 4th Cavalry and the 1st Engineers. In the resulting 2 hour battle, supported by Charlie Battery's fire, 73 North Vietnamese Army troops were killed. In the following week other convoys were attacked and an assault launched against FSB Thunder II. These attacks cost the enemy another 78 soldiers killed. During the "Battle of Thunder Road" a total of 151 enemy soldiers were killed. Just four days later it was our turn.

I spent almost half of my tour of duty at Fire Support Base Jim, and a lot of things happened there, many of them amusing. One soldier we called "The

---

[33] Mechanized.

Killer" took care of our pest problem of large rats! In Vietnam these hairy mammals often grew to the size of small dogs, much too imposing for a cat to handle. I once saw one that was leaning into a mess hall bucket eating garbage. His hind feet were on the ground and he was bent double, easily reaching the treasures at the bottom of the bucket!

"The Killer" stalked the exec post at night, with a flashlight and a machete. Deftly he checked the cracks in the walls of the bunker where the ammo boxes filled with dirt met and made perfect tunnels. With a lunge and a confirming rat's squeal he would add another victim to his "body count". Each morning he would have the fruits of his latest harvest laid out for the battery commander to see. The captain always nodded his thanks to "The Killer", and then would shake his head in wonder as he walked away.

There was a Vietnamese woman who did our laundry. Each day she made the trip from the small village to our south to wash, starch, and iron our uniforms. She had a young daughter who always came to me for the extra soap we got in SP (supplementary) packs. She would say "Docsan, mamasan send me get soap." and I would always oblige. If she or any of her family needed basic medical care I would give it. They were always grateful, especially for aspirin. I can't remember her name, if I ever knew it, I just called her Babysan. She gave me a Vietnamese "friendship bracelet" as a token of thanks. You can see it on my wrist in the photograph where I am sitting in the battery commander's jeep. Later I gave it to Carol's younger sister, Rhonda, who was about eight years old when I came back from the war. She still has the bracelet as well as the Vietnamese "Geisha" doll I bought her in Saigon all these many years later.

Other small incidents come to mind such as the Cobra that infested one of our latrines and rose up in the face of a GI as he sat doing his business. He cleared that outhouse fast I can tell you! Then there was the FNEWGY (fucking new guy) who cooked his rations over C4 and then stomped it out. He lost his rations in the resulting explosion as well as the bottom of his combat boot. Luckily he wasn't injured badly, just knocked unconscious.

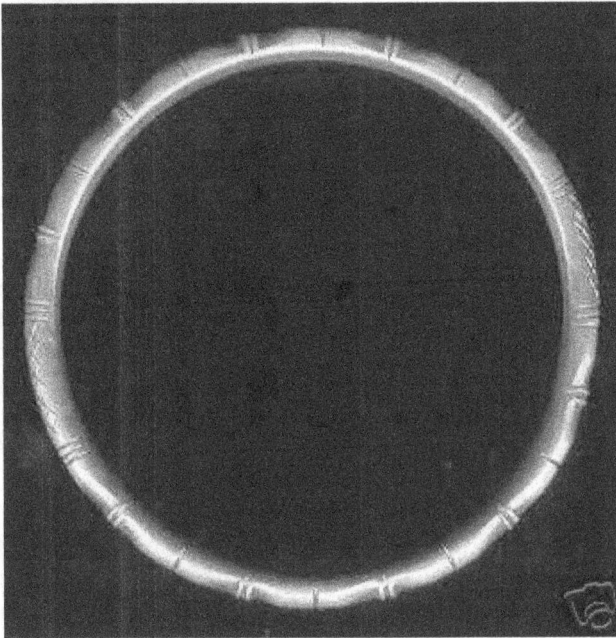

Vietnamese friendship bracelet, similar to the one I received from the laundry lady's daughter, Babysan.

Or there was the time enemy 122mm rockets came in and blew up our shower and water buffalo[34] and nothing else! The commander had requested a replacement for the leaking buffalo that had been . denied. He had a hard time convincing the battalion S-

---

[34] Water trailer.

4 officer that we really had lost the buffalo in "combat". Increasing the S-4's skepticism was the fact the water trailer was not blown up, but was smashed by a dud enemy rocket that failed to explode. However, a copy of the photograph I took and have depicted below finally convinced him.

The suspect water buffalo, photographed by the author.

In another instance that stands out in my mind, we had a skirmish with the enemy at night when I just happened to be on tower duty with a starlight scope and an M-60 machine gun. We killed our barber that night. Seems he cut our hair by day and was a VC by night! He and a small group of VC "sappers" tried to infiltrate our perimeter but were detected. We opened up with the M60 on several moving shadows and saw at least three tracer rounds pass through the enemy point man. Whether it was the barber we took out or one of the other members of his squad that was killed I don't know. But our intel guys told us he was our barber. It was a long time without haircuts for many of us before we got another barber.

I have often said that I must have filled a million sandbags while I was in Vietnam. Although an exaggeration, it certainly seemed like it. One sunny day while pursuing this task, I dug my heel into the ground to get good footing and saw something come up out of the sandy soil. At this point on the base several feet of dirt had already been removed to form the dirt berm around the base. Now I had dug down even further and there, at what I estimate was three feet lower than the surrounding terrain, I found an amulet. It was curiously light and tarnished with a green patina. On one side was Chinese characters, not Vietnamese, and on the other an elaborate dragon. It was one of the few things I brought back from Vietnam. Many years later I took it to an Asian scholar who told me it was

This is the 400+ year old amulet I found while digging to fill sand bags at Fire Support Base Jim. Note that the writing is Chinese rather than Vietnamese. Both the dragon and the symbols bestow good luck. The arrow points out the dragon's eye. I named it "The Dragon of Vietnam".

at least 400 years old. He went on to explain that it was Chinese silver that the dragon and symbols were for good luck and it was actually a form of money.

He indicated that when people felt threatened, money was often buried, and had I dug down a little deeper, I might have found a cache of Chinese silver, just my luck. However, this unusual item has become a family heirloom.

Hawaii

I can give you the exact dates I spent in Hawaii with Carol, because the day I returned to Vietnam is the day Apollo Eleven landed on the moon. That date is July 21, 1969. I had spent the previous seven days in the paradise of Hawaii, made even more of a paradise when compared to Vietnam. That means that Carol and I began the greatest vacation of our life on July 14th, 1969.

When my plane touched down and I, along with dozens of other soldiers, airmen, marines, and sailors got off, there was a long double line of wives and sweethearts waiting for us. I sauntered down the line in my khaki uniform all tanned a dark brown, my hair and mustache bleached almost blonde by the Southeast Asian sun. When I walked up to Carol she was busy looking around me for me. She did not recognize me until I spoke to her. My first words were "Hey good lookin', who you lookin' for?" She gave me a startled look and then burst out "Walt!" It was one of the best homecomings I ever had.

We spent that week getting to know one another again, exploring the beaches and mountains of Hawaii, and eating steak and eggs every morning for breakfast. At least I did. But there was a dark lining to the vacation sun of Hawaii and I got a taste of what I would experience for years to come. I found I was very uncomfortable around groups of people and I was constantly looking around and on my guard. A car backfired while Carol and I were

walking down a Honolulu street and I hit the pavement. The recognition of post traumatic stress disorder or PTSD[35] was still years away, but I already had it.

On Monday July 21, 1969 Apollo Eleven landed on the moon and Neil Armstrong was the first man to walk on the lunar surface. It was also the day I got back to Vietnam from Hawaii so I got to watch the landing on a black and white TV before I returned to the field. I later got to see and meet Neil Armstrong in person at the Bob Hope Christmas Show that December at the division's field headquarters in Lai Khe.

So while the astronauts were flying to the moon, Carol and I were flying to Hawaii. She came from the U.S. mainland, and I from Vietnam by way of the small Pacific island of Guam.

---

[35] Post traumatic stress disorder (PTSD) is an anxiety disorder that can develop after exposure to one or more terrifying events in which grave physical harm occurred or was threatened. It is a severe and ongoing emotional reaction to an extreme psychological trauma.

The Fight at Fire Support Base Jim September 9, 1969
A Final Act of Valor

Duty officer's log, Headquarters 1st Battalion, 7th Artillery 9 September, 1969 (declassified).

Entry number 4 at the early morning hour of 1:27am begins the dispassionate recording of the fight for Fire Support Jim with the entry "A 1/7th receiving incoming (82mm)."[36] The note that dusters spotted mortar tubes refers to a tracked vehicle armed with twin 40mm cannons. These vehicles were situated on the south side of the fire base and spotted the flash of mortar rounds as they exited the tube. Another entry

---

[36] 82 millimeter mortar, a high explosive round.

56

at 0139 hours reads "A, 1/7th to fire killer jr. everywhere except the northwest. Taking RPGs[37] through the wire." A little further down at 0135 hours is the chilling entry "1 man missing an arm."

Other pertinent and interesting entries followed on the next page of the duty log.

"0143 [hours] A 1/7 says FDC blown away & probably could not fire a mission at this time.

0148 Dustoff to arrive from NW in 10 min.

0152 A.O. [aerial observer] 50E due on station [above FSB Jim] 10 min.

0155 Req. E.O.M [request end of mission] on HE [high explosive rounds] from C 8/6 [Charlie Battery 8th Battalion, 6th Artillery] incoming stopped Hunter-Killer to work around perimeter.

0155 30 [LTC Sperow, battalion commander] wants arty to the east of hwy through FSB Jim – Hunter-Killer to work western ½ [of the base].

0158 Dynamite A 8/6 [Alpha Battery 8th Battalion, 6th Artillery, code name "Dynamite"] up and ready [for fire missions].

0200 Nighthawk[38] [surveillance helicopter] in area will go take a look.

---

[37] Rocket propelled grenades.

[38] To deny the enemy freedom of movement at night a "night fighter" helicopter was developed and equipped with night sensors. The sensors consisted of a mounted "NOD" (night optical device, or starlight scope) and a Xenon searchlight with both infrared and white light capability. Armament was an M-134 multi-barreled minigun. Nighthawk was usually escorted by one or two gunships, and flew between 500 and 1,000 feet above

<u>0210</u> Sit. Rep. [situation report] Sapper got through wire & hit FDC.

<u>0217</u> A.O. 50E out of Phu Loi.

<u>0218</u> A 1/7 to cut off Killer Jr. for Hunter-Killer team and Spooky to work over area. [A 1/7] EXP [expended] 368 HE 140 Self Illumination [rounds].

This was a lot of support for us that arrived very quickly. The attack on Charlie Battery 1/7th four days before had served as a wakeup call for local command operations.

Killer junior is a reference to a deadly anti-personnel shrapnel artillery round. It was commonly called a "beehive" round and was packed with 8,000 flechettes, a French word for "little arrows". It is called a beehive round because of the distinctive buzzing noise the darts make flying through the air. This ammunition is devastating when used against ground troops due to its shotgun effect. When used effectively, it can sweep enemy soldiers from the battlefield, or nail them to jungle trees.

I was awakened around midnight by the sound of incoming mortar rounds. At first I thought it was the guns firing a mission, but the flat "crump" sound to the explosions alerted me to the fact it was something else. A 105mm howitzer has a distinctive "ring" when it sends a round down its steel tube. I woke up the other two guys in my bunker, one the battery commander's driver and the other a member of the "exec post" or headquarters crew. I told them to lock and load and fire at the first thing that appeared in the doorway of our bunker. The VC were well known to

the surface.

send "Sappers", men armed with explosive satchel charges, that move about tossing them into bunkers.

Suddenly, just above the sound of the explosions I heard "Doc, Doc, I got a man at the FDC (fire direction control center) with his arm blown off."

It was the battery commander, Captain Harry G. Madden. In my haste I tossed my .45 on my cot and grabbed my aidbag. The only thing on my mind was what to do for a severed limb.

"Stay low, there's small arms fire and RPGs comin' in." The captain said as I exited the bunker.

I glanced to the left in the direction of the FDC and saw numerous explosions in that area. A three quarter ton truck was parked midway between my bunker and the first of the gun emplacements with their sandbag walls. I low-crawled to the truck wearing nothing but pants and a helmet, my boots left behind, and took cover under it.

I could hear the stutter of AK 47s and the crack of rounds passing close overhead. It reminded me of basic combat training when we had low-crawled through strung barbed wire while an M-60 machine gun fired over our heads and quarter sticks of dynamite exploded around us.

Scanning the area ahead of me, I crawled to the first gun emplacement and looked back for the captain but couldn't see him. About that time the truck I had just left exploded as an RPG (rocket propelled grenade) struck it and ignited its fuel. It began to burn brightly.

To escape its revealing light I moved quickly through the warren of sand bagged gun positions, stopping behind each position before moving ahead. Laying on the ground and peering around a corner I was suddenly lifted from the ground, my ears ringing as a hand grenade exploded just on the other side of the sandbag wall I was hiding behind.

"I've been spotted!" I thought and got to my feet, sprinting the last few yards and dodging around obstacles to the entrance of the FDC. As I arrived I saw him, a small VC soldier dressed in the usual black silk pajamas. He must have used up his last grenade, because he turned and ran. Seeing as how I didn't have a weapon, I was pretty glad he'd decided to do that. I stepped to the black opening of the FDC and yelled "Where's the wounded man?"

Receiving no answer I yelled it again. Still no one replied to my question. "Oh God, they're all dead." I thought, and moved into the bunker. The sharp stench of cordite, that telltale smell of a recent explosion, was so strong it burned my throat.

I pulled my red lensed flashlight from my aid bag and immediately saw a casualty at my feet. It was the soldier with the serious arm wound.[39] It wasn't completely severed, and I reached and pulled his arm into my lap.

A soldier named Mize, recently reassigned to us from the 82nd Airborne, held the flashlight for me as I applied a tourniquet. The wounded soldier, Don DeVore, who had just moved into the FDC the day before, lay on his back, his blood pumping out in long ropy strands to splash on my bare chest and arms. I adjusted the tourniquet, made from a medical cravat[40] and a stick. When the blood flow slowed and then stopped I was relieved. I dressed the wound and then secured his arm to his side. Two of the soldiers in the FDC helped me place him on a medical litter. I wrapped a second cravat around his wounded arm and moistened it with water from a canteen.

---

[39] Years later I learned this soldier was Donald E. DeVore when artilleryman Paul Jones saw DeVore give an interview on the History Channel in November, 2011.

[40] An olive drab colored scarf used for making splints for broken bones and tourniquets.

I turned to Marvin Millhouse, the radio operator, and asked for a MEDEVAC "dustoff" chopper. While he radioed battalion headquarters I pulled chunks of shrapnel from his back and dressed his dozens of wounds until he collapsed from the constricting shredded muscles. Marvin received the Army Commendation Medal with "V" for continuing to man the radio after being wounded. I would treat Marvin for months to come, removing pieces of metal as they surfaced and fighting back against any onset of infection.

Sergeant Bob Abbott of the FDC picked up the PRC-29 radio and helped maintain contact with both Division Artillery and the dustoff chopper. The VC had managed to place an anti-tank mine atop the FDC and destroyed most of the antennas. Bob had interviewed DeVore that day to fill an opening on the FDC crew. He had told DeVore to move into the FDC that night. When the fight began DeVore had hunkered down behind a footlocker but left his arm atop the locker. When the first hand grenade went off it blew up a metal ammunition box that immediately turned into the shrapnel that severed DeVore's arm.

I treated a couple more minor wounds as the sounds of the continuing battle came to us. Our battery was up all six guns firing direct fire high explosives and the devastating "beehive" round. These small but deadly missiles could nail a man to a tree, and often did.

Despite the incoming fire of mortars, RPGs and small arms, the dustoff was coming in. I organized a litter detail for the seriously wounded and led it out through the incoming fire and stopped them just inside the last sandbag parapet nearest the road which ran down the middle of the compound, the only place the chopper could set down.

Low-crawling to the bullet-whipped road I pulled out a battery operated strobe light, and waited for the chopper. While lying on my back I watched the tracers arcing over our position and sometimes disappearing into the sandbag parapets. Illumination rounds catapulted parachute flares into the darkness to float slowly down, their swinging motion and red light causing shadows to come alive and dance. The occasional hot streak of fire denoted the passing of a rocket-propelled grenade followed shortly by a shattering explosion.

The enemy had no idea I was there, but they would in a minute. Hearing the chopper over the blast of the guns, I turned the very bright white-light strobe on, and held it as high above my body as I could. The flashing light immediately drew enemy small arms fire and mortars began to fall around my position. But I could hear the chopper, and out of the darkness he came despite the tracers reaching for him and the fiery trail of an RPG as an enemy gunner tried to take the medivac down. He settled onto the road beside me, and as his door gunner fired into the enemy's position and our battery opened up with the hot steel of protective fire, I motioned the litter team forward and loaded our wounded consisting of Don DeVore and Marvin Millhouse onto the chopper. I got the impression of a red haired burly warrant officer pilot and then with a roar and a cloud of dust the chopper lifted away, taking those men to safety. I returned to the gun sections and as they continued to fight the enemy throughout the night, I moved among them treating small wounds.

One individual I approached, a very young buck sergeant, was wounded in his left shoulder. He refused medical attention, continuing to lead his soldiers in the defense of our position. Seeing that it was only a flesh wound I didn't insist. Little did I know that my remark to the commander later about him refusing treatment would help earn him the coveted Silver Star medal for gallantry in action. We fought all night, and as the morning sun rose, the enemy melted away, taking his dead and his wounded with him.

In the heat and confusion of the assault, the crew of the number 3 gun just outside the FDC bunker, had thrown a thermal grenade down the gun's tube to keep it from falling into the hands of the enemy.

Some time passed and then a runner came from the engineer platoon on the north side of the fire support base. The engineers had just arrived at Jim that same night. He told me they had a man down and would I come take a look. I followed him to where the platoon sergeant lay unmoving on the floor of the engineer platoon's tent. I played my light over his face it was contorted as if in pain. I got no reaction from his pupils, they were fixed and dilated. I felt for a pulse, but there was none. I looked him over and could find no wounds. It was clear he was dead and I informed the engineers. I inquired later on with the Divarty[41] surgeon about the middle aged platoon sergeant and he told me he had died of an apparent heart attack. Had I been notified earlier I might have been able to save him.

The battery commander called me to the exec post and wanted the names of the evacuated casualties and those to be put in for the Purple Heart for wounds received in action. I filled out the report and the commander wanted to know why my name wasn't on the list for the Purple Heart and he pointed out the many cuts my feet and legs had sustained from my movements during the battle. I thought a moment about all the men who had sustained serious wounds

---

[41] Division Artillery.

and the engineer sergeant who'd been killed and then declined. I just didn't think my wounds warranted a Purple Heart. I told him about the sergeant who refused treatment and he made a note of it. He later recommended the artillery sergeant and myself be awarded the Silver Star for "Gallantry in Action".

The Silver Star is the third highest decoration awarded for heroism. Needless to say I was pretty impressed at being recommended for it. But it was not to be. Back at Divarty HQ the recommendation was downgraded to a Bronze Star for Heroism (with a valor "V" device) in ground combat.

The artillery sergeant who I mentioned to the battery commander as refusing medical treatment received the only Silver Star awarded to a 7th Artillery soldier in 1969.

```
                      DEPARTMENT OF THE ARMY
                 Headquarters, 1st Infantry Division
                      APO San Francisco  96345

GENERAL ORDERS                                      15 October 1969
NUMBER   12917

                    AWARD OF THE BRONZE STAR MEDAL

1.  TC 320.  The following AWARD is announced.

CROSS, WALTER L  ████████  SPECIALIST FOUR United States Army
Battery A 1st Battalion 7th Artillery

Awarded:          Bronze Star Medal with "V" device
Date of action:   9 September 1969
Theater:          Republic of Vietnam
Reason:           For heroism not involving participation in aerial flight,
                  in connection with military operations against a hostile
                  force in the Republic of Vietnam:  On this date, Specialist
                  Cross was serving as a medical aidman with his unit when the
                  friendly encampment was suddenly subjected to an intense
                  rocket-propelled grenade barrage followed by a massive human
                  assault which caused numerous casualties.  With complete
                  disregard for his personal safety, Specialist Cross left
                  his relatively secure position and maneuvered through the
                  hostile fusillade to the aid of his fallen comrades.  After
                  administering first aid to the injured personnel, Specialist
                  Cross organized the prompt removal of the casualties to the
                  medical evacuation zone.  His courageous initiative and
                  selfless concern for the welfare of his fellow soldiers were
                  instrumental in saving several friendly lives.  Specialist
                  Cross' outstanding display of aggressiveness, devotion to
                  duty, and personal bravery is in keeping with the finest
                  traditions of the military service and reflects great credit
                  upon himself, the 1st Infantry Division, and the United
                  States Army.
Authority:        By direction of the President, under the provisions of
                  Executive Order 11046, 24 August 1962.

FOR THE COMMANDER:



OFFICIAL:                              A. G. HUME
                                       Colonel, GS
                                       Chief of Staff


J. P. BOTT
First Lieutenant, AGC
Assistant Adjutant General

DISTRIBUTION:
```

A copy of the original order typed in Vietnam. Notice the misspelling of the word "selfless". The black box conceals my social security number. A year before the use of the SSAN replaced the serial number issued each soldier upon enlistment or draft. If enlisted the number would begin with an RA designation for Regular Army followed by eight numbers. If drafted, the soldier's number would begin with US.

Captain Madden indicated to me I likely didn't get the Silver Star because I didn't tell him about the wounds to my feet and to tell him to put me in for the Purple Heart. He saw to it I received a second Bronze Star before I left, this one for meritorious service in a combat zone. Later in my career an army interview board wanted to know how in the world I had earned two bronze stars only a few days apart. I just shrugged. But the loss of the prestigious Silver Star was disappointing.

Lieutenant Jeffery who was in charge of the FDC at the time of the attack also got the Bronze Star with "V" for kicking two enemy hand grenades back out the bunker door, preventing even more injuries or possibly deaths. One of the enemy Chicom[42] grenades we found in the FDC bunker the following morning had not gone off. Charley had pulled the pin, but forgot to remove the rubber band wrapped around the charging handle, preventing its detonation.

Many years later on Veterans Day, 2011 I received an email from 7th Artillery veteran Paul Jones. He told me that he had seen Don DeVore on the History Channel telling of his wound in Vietnam and that he was sure this was the soldier with the arm wound I treated. I found an article online about Don written by *New Jersey Herald* reporter Jessica L. Mickley and dated in early November, 2011.

A History Channel miniseries featuring [Don] *DeVore and 12 others affected by the Vietnam War...premiered on Tuesday* [November 8] *... "I'm flattered, but at the same time, it's a little overwhelming," DeVore said...Thirty-five years after serving in Vietnam, DeVore sought treatment for what he now knows is post traumatic stress disorder. He attends a weekly group therapy session at Newton Medical Center with other veterans.*

---

[42] Chinese Communist.

DeVore's tour of Vietnam lasted seven months. Midway through DeVore was sent home to be with his pregnant wife, who doctors thought might have complications during labor. He was in transit when his daughter was born, possibly somewhere over the Pacific Ocean, but he can't be sure. DeVore arrived on the third day of his daughter's life, just in time to usher his wife and baby girl home from the hospital.

DeVore had great timing, however, for the music event of a lifetime. He rounded up a few friends and went to Woodstock to see the Who.

"I think I was the only person there with short hair," he said. Returning to Vietnam afterward was "Dreadful," DeVore said. "That was the worst trip of my life."

Five weeks after landing in Vietnam, DeVore was severely injured by a rocket-propelled grenade [actually it was a hand grenade], *and spent the next two years of his life in the hospital. Scars from numerous surgeries remain as an unnecessary reminder.*

Nowadays DeVore is doing "All right." He does not love the attention resulting from his History Channel appearance (his children keep reading posts about him on Facebook). He's just one of many Vietnam veterans, he says, with many unique stories. DeVore participated in the documentary because he admired the program's intent in presenting Vietnam through personal stories. "It's not blood and guts," he said. "It's what people really went through."

In a related article the same reporter wrote the following also regarding Donald DeVore.

Some veterans, like those who served in Vietnam, receive gratitude now, though that was not always the case. These veterans not only fought for their country and their lives, but came back to the states to fight an army of backlash.[43]

"We were baby killers. We were all drug addicts," Vietnam veteran Don DeVore of Franklin [New Jersey], *said.*

DeVore, a regular attendee of the county Veterans Day celebration, was wounded by a rocket propelled grenade about seven months into his tour. He spent the next two years in the hospital, in and out of surgeries. He came out with an arm of scars and a lot of shame.

"There were times I wouldn't even acknowledge that I was a veteran," DeVore said.

For a while, when asked about his scars, DeVore would blame a motorcycle accident, until about 16 years ago. His then 16-year-old daughter entered an essay contest "What Democracy Means to Me", through the VFW.

"It's the first time it even really dawned on me that my children were really proud of my service," DeVore said.

The History Channel video DeVore appeared in is titled *Vietnam in HD* and the producer wrote a synopsis of DeVore's story.
Monticello, New York
U.S. Army, 1ˢᵗ Division. Service: Spring 1969 – Summer 1969.

Like thousands of other young American men, Don DeVore struggled intensely with what he would do if he were drafted to serve in Vietnam. He had no desire to become a war hero, and no dreams of winning glory or greatness on a battlefield.

---

[43] See my book *Fighting the 2ⁿᵈ Vietnam War at Oklahoma State University.*

In the late summer of 1968, DeVore's number was called and within weeks he was shipped off to basic training at Fort Jackson, South Carolina. Arriving in Vietnam in March of 1969, DeVore was assigned to an artillery unit at a fire support base known as Firebase Jim. His job was to provide accurate fire support for the search and destroy patrols that were taking place on a near daily basis in the surrounding jungles. After four months, DeVore was granted compassionate leave to attend the birth of his first child. Upon returning home, he found himself in the middle of the largest peace and love festival of the decade – Woodstock. It was a stark contrast to the harsh combat he returned to just days later. In September of 1969, the Viet Cong infiltrated Firebase Jim and DeVore was severely wounded by an RPG [sic] rocket propelled grenade), sustaining an injury to his left arm that kept him hospitalized for nearly two years. The psychological and physical effects of his combat experience were devastating. For years, DeVore never spoke about the war. When questioned about the scars on his arm, he would tell people they were the result of a motorcycle accident. Finally, in the late 1990s, he sought treatment at a VA hospital, and after several years of counseling he was finally able to come to terms with [his] wartime experience.

In November of 2011 I found Don's address online and wrote him a short letter. A few days before Christmas, on December the 21st, I received an answer back from him. It had been 42 years since we had last spoken to one another.

Doc Cross! Holy s**t. Man, it was great hearing from you! Before I go any further, please let me say "thank you" a thousand times for taking care of me! I know what you did – I read your book quite a while *ago. I wish you had written it sooner! I owe you brother! When I first heard about your book (and you, obviously), I Googled around and found you at OSU and tried to reach you. When I didn't hear back, I assumed you either were no longer at the school, or perhaps, you just weren't up to dredging up the past. Completely understandable! I know.*

I am doing very well. I'm in good health, have a wonderful wife, six (6) great kids, and four (4) beautiful grandchildren. I've pursued a career as a professional engineer and land surveyor and currently work for the Base Ops contractor at a large, believe it or not, Army R&D installation near my home (Picatinny Arsenal, NJ).

Please, please, let's stay in touch. God bless you Walt. Thanks so much for reaching out to me! I'm looking forward to hearing back from you.

Don DeVore Franklin, NJ

His letter was a great Christmas gift.

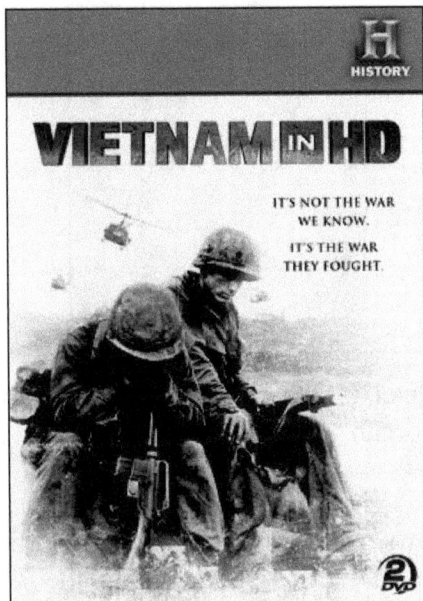

Vietnam in HD, known as "Vietnam Lost Films" in countries other than the U.S., is a six part American documentary series originally shown on the History Channel in November of 2011. It remains offered for sale as a 2 DVD set. It is narrated by Michael C. Hall, known to many of his fans as Dexter Morgan from the *Showtime* original TV series *Dexter*.

Actor Michael C. Hall, better known at the time as Dexter Morgan in 2011. Photograph by Keith McDuffee

What follows is a synopsis of that portion of the Vietnam in HD series pertaining to the 7th Artillery that appears on the second DVD.

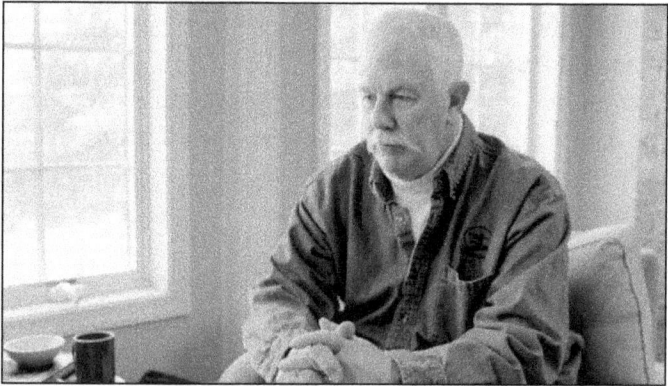

Don Devore as he appears in the documentary film Vietnam in HD.

A Changing War 1969 – 1970

Three minutes into the second DVD appears the following title in white lettering: Fire Support Base Jim Spring 1969. The film footage is stock; the scenes depicted are not from the actual Fire Support Base Jim. Former Battery A, 1/7th Artilleryman Don Devore is introduced as an artilleryman of the First Infantry Division stationed at FSB Jim. There is no actual mention of Alpha Battery or the 1/7th Artillery. The narrator describes Devore's background and he is shown as he himself picks up the narrative. The film then goes on to other Vietnam vignettes, and returns to Devore a few minutes later.

Ten minutes into the documentary it returns to Donald Devore who continues to narrate his return home on emergency leave and attending the festival at Woodstock.[44] Don had left Vietnam three days before the anti-Vietnam War festival. What a change of venue that must have been for him!

Once again the title returns as Fire Support Base Jim and more stock footage of artillerymen is shown. Devore mentions the death of Ho Chi Minh on September 9, 1969 and the declared truce for that night. Don narrates the events of the attack as he remembers them.

Don had lost a lot of blood and was in shock which likely colors his memories and his story when he states near the end of the film sequence that:

---

[44] The Woodstock Music & Art Fair, informally known as the Woodstock Festival or simply Woodstock, was a music festival, billed as *An Aquarian Exposition: 3 Days of Peace & Music*. It was held at Max Yasgur's 600-acre dairy farm in the Catskills near the hamlet of White Lake in the town of Bethel, New York, from August 15 to 18, 1969. It was an anti-Vietnam war event.

"When I opened my eyes a bunch of guys were cheering for me, kind of like a football player who has been hurt on the field now being carried off to the yells and the screams of the crowd. As we lift off [in the medivac chopper] *a calm starts to come over me. I can see out the side door that the morning sun is starting to set in with a purpulish* [sic] *red sky. All I can think to myself is I'm alive! For me, this craziness is over."*

With those words the sequence about Fire Support Base Jim ends. After pondering his words and impressions I can draw a few conclusions from them. I had not administered morphine to Don at the time he was wounded, a fact I imparted to the medic on the helicopter. Without a doubt the calm that came over him was from morphine given him by that medivac medic. As for the dawn statement, the helicopter arrived at 0148 hours, or a little before 2 a.m. in the morning. We were many hours away from dawn and his flight onboard the chopper was likely no more than twenty to thirty minutes in duration either to the division's forward operations base at Lai Khe, or to Saigon. Don's memory of a morning dawn is a false memory and may also be attributed to the morphine he received on the helicopter.

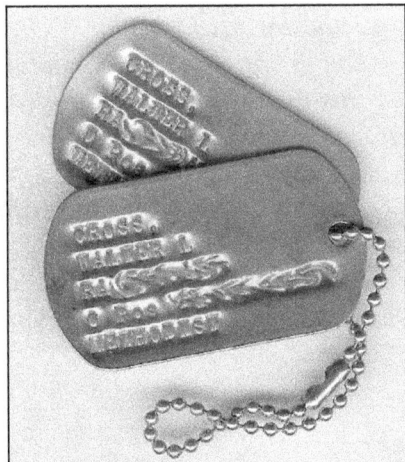

The dog tags I wore in Vietnam. Both my RA number and my SSAN were displayed on them.

CITATION

BY DIRECTION OF THE PRESIDENT
THE BRONZE STAR MEDAL
FIRST OAK LEAF CLUSTER
IS PRESENTED TO

SPECIALIST FIVE WALTER L. CROSS,

HEADQUARTERS HEADQUARTERS AND SERVICE BATTERY, 1ST BATTALION, 7TH ARTILLERY

1ST INFANTRY DIVISION

who distinguished himself by outstandingly meritorious service in connection with military operations against a hostile force in the Republic of Vietnam. During the period December 1968 to December 1969

he consistently manifested exemplary professionalism and initiative in obtaining outstanding results. His rapid assessment and solution of numerous problems inherent in a counterinsurgency environment greatly enhanced the allied effectiveness against a determined and aggressive enemy. Despite many adversities, he invariably performed his duties in a resolute and efficient manner. Energetically applying his sound judgment and extensive knowledge, he has contributed materially to the successful accomplishment of the United States mission in the Republic of Vietnam. His loyalty, diligence and devotion to duty were in keeping with the highest traditions of the military service and reflect great credit upon himself and the United States Army.

This is the citation for award of the 2nd Bronze Star. In awards for service the citation and order are separate unlike the award for valor. The order was dated in November of 1969, so Captain Madden recommended this award for me right after I received the first award. He wanted me to receive this second Bronze Star before I left country. It was presented to me by the battalion adjutant in his office at the forward headquarters base of Lai Khe just before I got in the jeep for departure. There is a saying among Vietnam veterans, "We were winning when I left..."

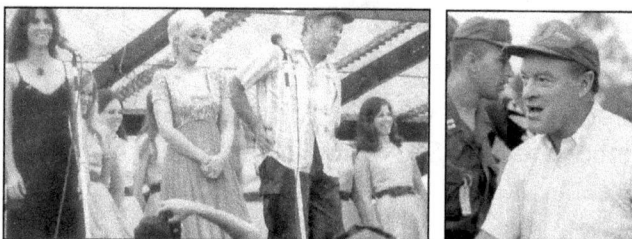

In December of 1969 I got to see the Bob Hope Christmas show when I flew on a Huey from FSB Florida to Lai Khe. Note the 1st Infantry Division patch on Bob's hat. That is Connie Stevens beside him, also I got to see and meet Astronaut Neil Armstrong just back from the moon!

My award of the Bronze Star came toward the end of the assignment of the 1/7th Artillery in Vietnam. Along with the other men's decorations, it was one of the last awards for heroism earned in a final act of valor by 7th Artillery members just before the end of the unit's history in Vietnam.

The fight for Fire Support Base Jim was not yet over. Compared to the early morning hour's action, the second attack was not as intense and this paragraph from the unpublished history of the 7th Artillery sums it up as follows:

Evidently assuming the personnel at the fire support base dropped their guard because of the previous early morning activity, the enemy again attempted to breach the perimeter a second time that night. This time, however, the ground surveillance radar[45] set at the fire support base detected the activity and the enemy was engaged with small arms and "Killer Junior.

Heavy blood trails indicated the enemy had suffered casualties again. No friendly forces were wounded during this second attack.

---

[45] Ground surveillance radar is ground-to-ground surveillance radar set for use by military units. The radar is capable of detecting and locating moving personnel and vehicles.

A Summary of 7th Artillery Operations in Vietnam

Operations continued without letup despite the attacks on Charlie and Alpha batteries. A combined task force of US and ARVN units sealed off and searched the village of Phu Hoa Dong. Long a hub of Viet Cong activity, the village came under siege with Bravo Battery firing out of Fire Support Base Mortain, delivering a total of 1700 rounds of high explosive artillery. The siege was maintained by infantry of the 2nd Brigade and the 7th ARVN Regional and Popular Forces. The enemy lost twenty-three dead, 17 captured VC soldiers, and 16 that rallied to the government of South Vietnam. There were minimal friendly casualties.

On October 1, Bravo Battery moved from FSB Mortain to a temporary stay at Di An base camp before moving on to FSB Florida. FSB Florida was destined to be my last position before leaving for "the world". There, Bravo Battery provided support to the 2nd Battalion, 18th Infantry.

By the 10th of October all firing batteries of the 7th Artillery were positioned for a planned artillery preparation except for Charlie Battery. To support and augment the 7th Artillery's fires, Bravo Battery of the 8th Battalion, 6th Artillery was displaced from Thunder III to Lai Khe while Charlie Battery, 1/7th occupied an area north of Dog Leg Village.

All assigned units made radio contact with 1/7[th]Tactical Operations Center (TOC) at 0500 hours and began firing the TOT (time on target) at 0530 hours. The most distant guns would fire first so that all the shells would land on the enemy positions at nearly the same time. This coordinated fire would deprive the enemy of time to disperse and seek shelter from the rain of high explosives.

An 8 inch self-propelled howitzer, Bravo Battery, 7th Battalion, 15th Artillery in Vietnam 1969.

The TOT preparation consisted of three areas 400 meters in diameter and a stream bed approximately 1000 meters in length. The area was a known enemy concentration site.

Among the units involved in the preparation were on 8" howitzer[46] battery employed along the stream bed in an effort to inflict casualties within 100 meters of the watercourse; as well as a 175mm battery committed in a reinforcing role in the northernmost target area.

Alpha Battery participated in the TOT and I spent a good deal of time traveling back and forth between the guns, the fire direction control center, and the exec post. I didn't want to miss any of this event the first TOT I had witnessed. We could hear the distant low thunder of the heavy guns as they initiated the surprise barrage.

---

[46] These are some of the largest land based guns, three inches bigger than the five inch guns found on a navy destroyer, and half the size of the sixteen inch guns found on the largest battleships.

The awesome event and it was an event, especially for the VC, concentrated in the targeted area, lasted less than half an hour. Six artillery batteries of 105mm, 175mm and 8 inch guns brought fire and brimstone from the sky to land on Mr. Charles just as he was forming up for his day's operations. In the dry phrase of the unit history;

"The target area was neutralized by a total of 1,345 artillery rounds fired within a 20 minute time frame."

A rate of fire of more than 67 rounds per minute of accurate, high explosive artillery rounds. It must have seemed as though the world had come to an end for the Viet Cong despite their well-dug-in positions.

Our aerial observer said that whole trees were thrown into the air and the jungle set ablaze with the mixture of high explosive rounds and phosphorus shells. The darkness of black smoke was illuminated from within with each red explosion or shower of white phosphorus burst anew with a continuous deafening roar. It was the heaviest most devastating bombardment he had seen outside a B52 bomber attack. It dealt the Viet Cong and their supporting NVA regulars a crippling blow with lasting effects.

Thirty days later on October 31, 1969 the "Vietnam Summer-Fall 1969 Campaign" came to an end. I was about to earn the fourth and last campaign star to my Vietnam Service Medal during the following "Vietnam Winter-Spring Campaign" of 1969 - 1970.

In early November Phase III of Operation Toan-Thang terminated and Phase IV began. Charlie Battery moved from FSB Venable Heights to FSB Thunder I. On November 20 my unit exchanged places with Bravo battery and Alpha took possession of Fire Support Base Florida.

Also in November the name of a small Vietnamese village, My Lai, and the terrible things that happened there would burn into the minds of the American public. It was only now, late in 1969 that the massacre of innocent Vietnamese by American soldiers became known. The cover-up that had lasted more than a year was discovered. It would go down in history as "The My Lai Massacre."

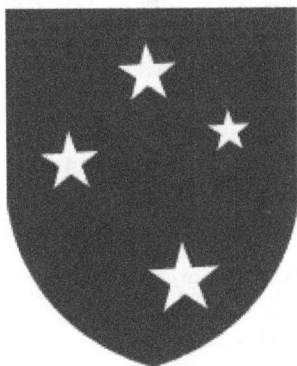

Shoulder sleeve insignia of the 23rd Infantry "Americal" Division. The patch is blue with white stars.

Besides the My Lai incident, the 23rd Infantry Division was further embarrassed in March 1971 when a company of the 196th Light Infantry Brigade was overrun during the "Battle of Fire Support Base Mary Ann". The unit suffered 33 men killed in action and 83 wounded.

As a consequence both the brigade and division commanders were relieved of command by July of that year. In November two of the division's three brigades were withdrawn from Vietnam and the division was deactivated. In what seems an odd decision, the 196th Light Infantry Brigade remained in Vietnam as a separate unit and was the last major combat unit to be withdrawn from the war.

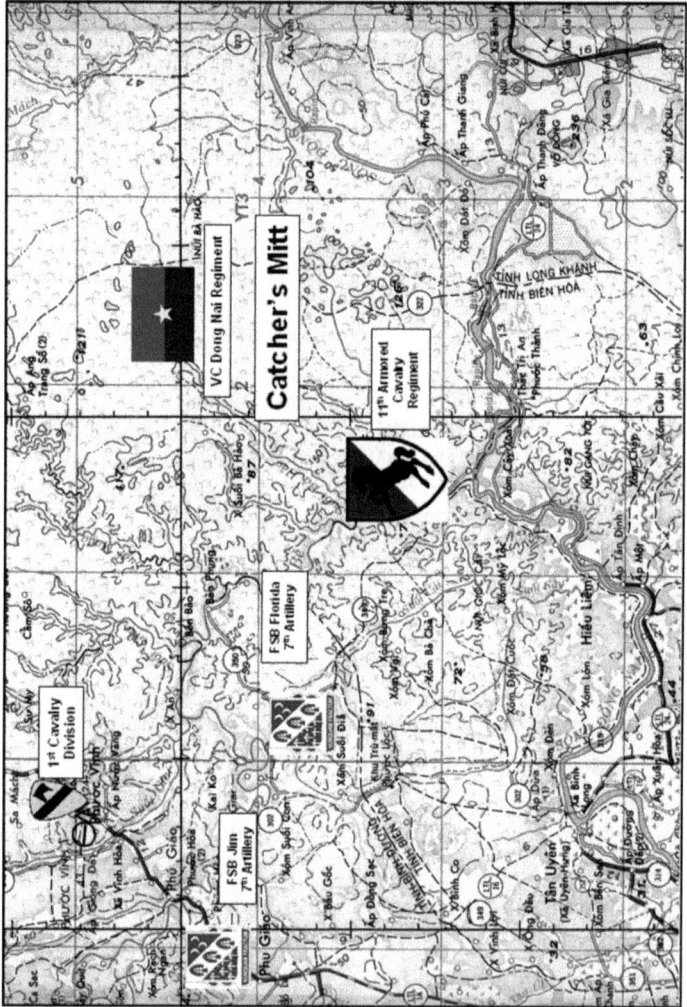

War Zone D showing the location of FSB Jim and FSB Florida.

Fire Support Base Florida

FSB Florida was located a short 3 kilometers east from the village of Khu Tru Mat and 20 Kilometers north and northeast of Bien Hoa airbase between Bien Hoa and Phuc Vinh. This area of War Zone D was known as the "Catcher's Mitt". The terrain was dense jungle bordering the western edge of foothills dubbed the Gang Toi Hills that rise to the east until reaching the Central Highlands.

This area of thick jungle is characterized by the Song Be River that twists and bends around the hills. Rubber plantations surround the small hamlets like Khu Tru Mat and consist of the homes of plantation workers and farmers. Following paths along the river is a portion of the Ho Chi Minh Trail winding its way toward Bien Hoa and Saigon. The 11th ACR (Armored Cavalry Regiment) under Colonel George S. Patton[47] was responsible for interdicting the enemy as they moved down the trail. It was down this same corridor that men and supplies for the Tet Offensive traveled both in 1968 and 1969. And I have no doubt that when we were moved into this area in December it was to be prepared to help stop movement along this area for the upcoming Tet holiday of 1970.

At FSB Florida we could see the "freedom bird" fly over us each day, taking lucky American soldiers, now Vietnam veterans, home to "the real world". When it came my turn to go home I looked down, and for the last time I could see FSB Florida and Alpha Battery. It was very strange indeed to see them from that vantage point.

For some reason, while at FSB Florida I never took any pictures. And yet at this base I met my first

---

[47] Son of the World War II general.

Red Cross girl, also called a doughnut dolly. Three of them flew to our base with the day's noon time hot meal and spent several hours with us.

Those meals, flown out onboard a slick and served from insulated mermite containers, always consisted of the same food.

Roast beef, mashed potatoes with brown gravy, green beans and dinner rolls. Good thing I loved those very food items, and looked forward to lunch every day. I never grew tired of it. The other two meals of the day consisted of "C" (combat) rations. I had my favorites among them as well. The canned pound cake and cinnamon rolls were great with morning coffee made from the pouch of instant powder found in each ration. I even liked the lima beans and ham, not usually a favorite among the GIs.

It was from FSB Florida I would catch that same lunch-time chopper in to Lai Khe to see the Bob Hope Show. It was about mid December, 1969 and when the chopper pilot offered to take passengers to see the show, I hopped on board. I always admired Bob Hope and to get to see him in person was a real treat. Not to mention Connie Stevens, the Golddiggers,[48] and the first man on the moon, Astronaut Neil Armstrong. It was a fine show and a welcome break from the jungle.

It was at FSB Florida that two odd things happened. The first was an Air Force event. A 5000 pound "bunker buster" bomb was dropped not far from FSB Florida. The problem was that it didn't go off. Fearful the VC would remove the explosives and use it against American troops, a second 5000

---

[48] A song and dance troupe of young women. The group did these goodwill tours for several years during the '60s, each Christmastime event lasting two weeks with Bob Hope and the Golddiggers doing two shows a day, closing with a version of "Silent Night" that reportedly reduced most of the girls to tears. The battlefields of Vietnam were a long way from Hollywood.

pound bomb was dropped on top of the first. We were ordered to take cover, but of course we watched. When the second bomb went off it caused the first to explode as well. The resulting mushroom cloud looked like a nuclear explosion rising into the sky. It shook us like an earth tremor.

The other event was more strange and unbelievable. FSB Florida was in the middle of the jungle. We got there by chopper, slicks carrying in the men and Chinooks carrying in our guns. We had no infantry with us, but the position was a strong one. Each gun position had a strong parapet built around it. In front of each parapet were bunkers with firing slits and an M60 machine gun in every other bunker. Concertina wire surrounded the position and claymore mines faced the jungle. In pits between the mines were tanks of Foogas ready for any enemy assault.[49]

At night Charlie would move about in the tree line and set up mortars to fire at us. They used small flashlights while setting up to fire and to counter them we used a starlight scope[50] to scan the woods at night. This device only needed the light of the stars to see into the darkest night. The scope showed everything in shades of green against darkness. While using the scope I noticed a cigar shaped object flying just above the tree-line.

---

[49] Foogas is a mixture of gasoline or jet fuel and powdered detergent in a 55 gallon drum buried outside the berm but inside the wire entanglements of a defensive position. it usually was activated by an electrical ignition attached to a C-4 charge

[50] Night vision devices used the principal of light intensification for vision. These were the first generation and the one our battery had cost $20,000 a large sum at the time.

This first generation of starlight scope cost $20,000 when the battalion obtained them in April of 1969.

"Enemy Helicopters"[51]
At Fire Support Base Florida

On December 29, 1969 I was nearing the end of my tour of duty in Vietnam

The battery was under the command of Captain Harry G. Madden, a young courageous artillery officer who assumed command of the battery in July or August of that same year. As a battery commander's tenure was usually six months, Captain Madden remained with the unit until it was send home three months later in March of 1970.

I had been with the battery for my entire twelve months of deployment in Vietnam, and although a medical aidman, I had learned many of the tasks of an artilleryman and was often pressed in to help out during fire missions. I enjoyed a certain amount of rapport with the captain who quite recently had

---

[51] UFOs were not reported as such in the official records of the Vietnam War, but were always referred to as "Enemy Helicopters".

87

recommended me for award of the Silver Star for Gallantry in Action stemming from a recent fight known as the "Battle of FSB Jim".

The position we occupied, FSB Florida, was in a somewhat secluded area known as the "Catcher's Mitt" for its local topography of hills surrounding a valley. In the "mitt" part of the valley ran a section of the Ho Chi Minh Trail leading to the northern approaches to Saigon only about 30 kilometers south. The area was the operational domain of the 11th Armored Cavalry Regiment (11th ACR) as well as the Dong Nai Regiment of the Viet Cong.

FSB Florida had no roads leading to or out of it. We were brought in by helicopter with our 105mm howitzers carried in one at a time by Jolly Green Giant transport choppers. While coming in the perimeter of the hill top position was secured by a heavy platoon of infantry. After we set up the guns and became operational the infantry would be taken out.

We had a good open view of the valley thanks to a defoliated jungle that reached out a good hundred yards around the fire base. Scattered here and there stood the naked trunks of trees we would use as target practice for our cannons in the coming days. This cleared the area even more, denying the VC or NVA any safe hiding places.[52]

A couple of nights into our stay I took a turn at perimeter guard from atop the exec post, a bunker that housed the headquarters personnel. The enemy often used flashlights to set up their mortars in the dark of the night, made even darker by the cover of the jungle canopy. To help spot these nighttime operations and bring our guns into play before the enemy could

---

[52] The Viet Cong and North Vietnamese Army respectively.

launch an attack, the battery had recently received a starlight scope.

The scope was excellent for picking up any light source as well as amplifying any available ambient light, such as the faint light of the stars. As soon as it got dark I began to slowly scan the wood line seeking the telltale flash of an enemy light. But my eye was brought up when I caught the glow of something above the trees. I brought the scope up and there, south of our position and seemingly moving quite slow, appeared a flying disc-shaped object in the familiar green color projected by the starlight scope. It was flying level above the tree canopy and moving, relative to my position, right to left or west to east. I did not automatically think of it as a UFO, in fact I thought it must be one of our helicopters despite its distinctly disc-shaped fuselage.

I visually tracked it with the scope as it turned in a wide arc and changed its direction of travel from east to north east and then north. Removing my eye from the scope I tried to make out the shape with my naked eye, but all I saw was the pitch black of the jungle. Although there were stars out and I watched for a shadow moving across them, I saw nothing. I also listened intently but could not detect the familiar "whump, whump, whump" sound of a helicopter.

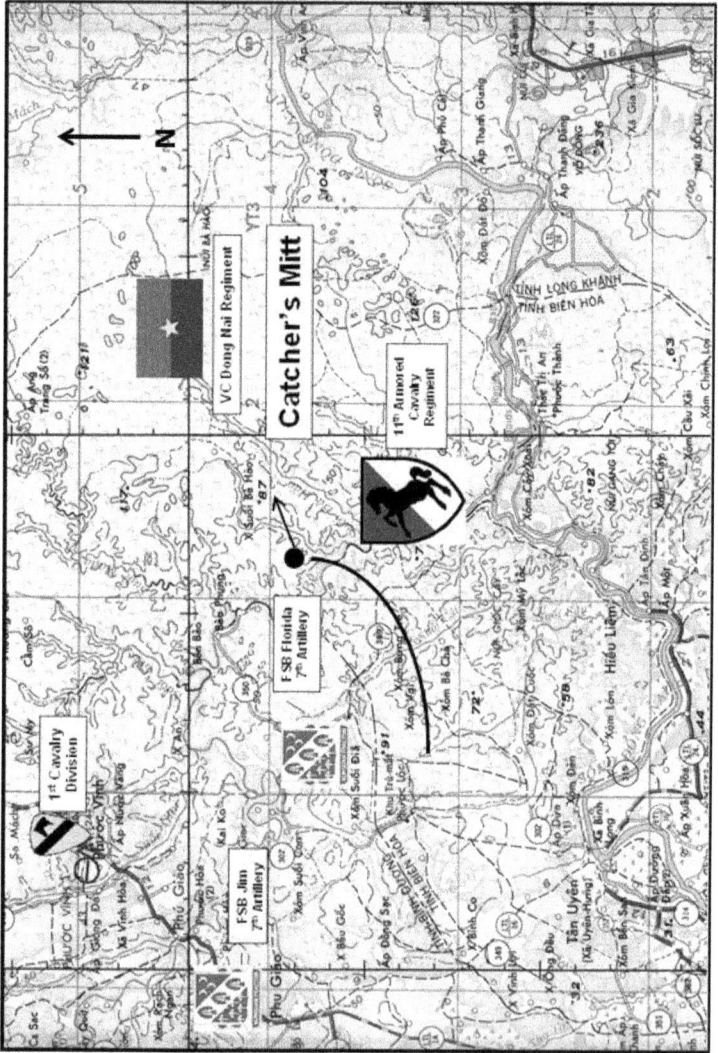

The first flight of the "enemy helicopter" is indicated by the arc. The second flight is shown by the arrow. Note the position of the 1st Cavalry base camp at Phuoc Vinh where the intruder was also viewed.

I returned my eye to the scope and it took me a moment to again gain a visual of the object. It had moved further north and as I continued to watch a second green lighted object appeared below the first. But this one was not flying and in fact was not a disc but a dome, and it sat on the southern slope of a hill easily seen in the daylight to our north east. What really caught my eye was the fact that the dome, although stationary, was pulsing slowly. The pulses seemed to come every five or six seconds, and I could see the disc tip its "nose" in relation to the dome and continue to move toward it at an angle, as though it were going to pierce the dome. I turned to one of the exec post personnel and said "Go get the captain, I got something." Then I returned to the scope and watched incredulously as the disc did indeed, pierce the dome and merged with it, slowly disappearing from view.

That is when Captain Madden showed up. "Whatcha got Doc?" he asked. I replied that I wasn't sure and invited him to take a turn with the tripod mounted scope.

"What the hell is that?" He continued. I told him what I had seen, the slow flying disc and the fact that it disappeared into the dome. "Yeah, I see it. It appears to have launched from the dome, or maybe it's a different disc. It's heading away north and east." Then he lifted his view from the scope.

"What do you think sir?" I asked. The captain just shook his head. "I don't know what to think. It doesn't appear to be a threat to us so ignore it and concentrate on watching for VC." He concluded and had started to walk away but then turned. "And I don't want any report on that appearing on the exec post log!"

I acknowledged his instructions and after looking at the dome for awhile longer I went back to searching the tree line. After making a complete sweep of our perimeter I went back to where the dome was positioned, but it was gone as if it had never been there.

About thirty kilometers north of Fire Support Base Florida was the base camp of the 1st Cavalry Division. Among the soldiers stationed there was a hometown friend of mine, Specialist David Lemon. David and I grew up in the same small town in Oklahoma and enlisted in the Army together on the "Buddy" plan. David was a military policeman and worked on the main gate of the base. Around the time I spotted the "enemy helicopter" to the south and east of the base a perimeter guard at the base also saw an intruder, perhaps the same intruder I saw. The name of the witness was not given in the report it may have even been David.

I never got to swap Vietnam stories with David, I was sent to Europe after leaving Southeast Asia while David went home, and died not long after. That report from Phuoc Vinh is presented and offered as a supporting statement to the Fire Support Base Florida narrative. When last seen by the 1st Cavalry Division soldier the intruder was seen moving south, in the general direction of Alpha Battery, 7th Artillery.

"Enemy Helicopter" Encounter at Phuoc Vinh North of FSB Florida

"I was pulling Green-line duty with 3 other 1st Cav. soldiers who were sleeping. I had a starlight scope, a radio and all the stuff you would expect in/on a bunker. This bunker was a big well-fortified bunker. We were all on top of this bunker from my best recollection. I was pulling my stint, letting the others sleep. This bunker was on the western facing perimeter. This night was a beautiful night with no overcast. Many small brilliant stars were in the night sky. No moon, as far as I remember especially facing out west. The starlight scope was working very well. I'm going into a little detail to set the stage leading up to my sighting. I know it's not to the point please bear with me.

As I was scanning the western night sky, all of a sudden something to the Northwest caught my eye. It was a very brilliant whitish, silver and with a hint of blue, more of a rounded shape. It was fairly far away. The main thing about this object was it would move to my left, or south in jerky movements, hover, do it again and again. It never lost the same brilliance or colors the entire time. Additionally it left an amber or reddish trail (like a tracer) as it moved only to suddenly stop on a dime. I watched this thing for several minutes.[53]

By this time it was in the southern horizon. Then, all of a sudden it shot up skyward on a 45 degree angle towards the north. It was totally out of sight in seconds. I was thinking should I report this? I decided not to. I would wait to see if anyone else would, then I would to. I didn't have the presence of mind to wake the other guys. Mainly, I couldn't believe what I was seeing."

While we were at FSB Florida the battalion tactical operations center moved to Lai Khe on December the 9th to support 2nd Brigade operations. The brigade was moving into a new area of operations north of Lai Khe between Thunder Road and the Song Be River located to the northeast of the big base camp.

On December 12 Bravo Battery moved its six guns from FSB Jim to FSB Oklahoma in support of an operation by the 1st Battalion, 26th Infantry along the western edge of the Song Be River. I have no

---

[53] He had to be viewing this object with the naked eye because a starlight scope shows all light images as green.

idea what happened to FSB Jim after the 7th left, but it was likely stripped of all useable military materiel, deconstructed, and abandoned.

Stand Down

Toward the end of December, it was announced by MACV[54] that the 1st Infantry Division was to be one of the units withdrawn from Vietnam as part of a troop reduction. This surprise announcement caused a shuffle of tactical areas of responsibility and the 2nd Brigade was assigned an area halfway between Saigon and the in-country R&R center at Vung Tau. The 1/7th TOC again moved, this time to Bearcat and the rear elements went back to Di An. This was the last move of the battalion in 1969.

On the first day of January, 1970 Charlie Battery moved from FSB Minnesota to Di An and the day after to FSB Dakota. On 3 January Alpha Battery moved from FSB Florida to Di An. On 5 January Bravo Battery moved from FSB Oklahoma to Di An and Alpha from Di An to the Australian base camp at Nui Dat for one day and then on to FSB Colorado.

The Fire Support Bases of Dakota, Rhode Island, and Colorado were the last three main fire support bases from which the maneuver battalions of the 2nd Brigade operated.

On January 19 Charlie Battery displaced three guns to a location 12 kilometers south of FSB Dakota near the village of Op Trinh. Naturally, this new location became known as FSB South Dakota and the old FSB became North Dakota. These four locations were the last positions of the 105mm firing batteries of the 1st Battalion, 7th Artillery, known as the Pheons.

The battalion had again demonstrated its worthiness of its motto "Never Broken" sometimes spoken in the Latin phrase "Nunquam Fractum".

---

[54] Military Assistance Command, Vietnam.

On March 1, 1970 Alpha Battery moved from FSB Rhode Island to Di An to counter enemy activity in the surrounding area and maintained its operational status beyond the projected date for stand-down. The very next day Bravo Battery came in to Di An and began to stand-down. On March 3 the last of the firing batteries, Charley Battery, made their way to Di An and also began to stand-down. That night, Alpha Battery terminated operations, the last of the battalion to do so, bringing an end to the Vietnam era of the 1st Battalion, 7th Artillery.

The "Pheons" had done their job, firing 1,250,000 rounds of artillery fire that no doubt saved as many friendly lives as it destroyed in enemy casualties. During combat operations in 1969 nearly 240,000 rounds of 105mm ammunition was fired, inflicting 359 enemy killed, thousands wounded, and resulting in 33 Viet Cong and North Vietnam Army soldiers to "Chu Hoi".[55]

At the battalion awards ceremony on 4 March 1970, Brigadier General Wolfe, the 1st Division's assistant division commander for maneuver elements commended the unit for its meritorious service. The Vietnam service of the Pheons was over. Little did we know that the Pheons themselves would soon become a thing of the past.

---

[55] Surrender. Literally translates into "open arm".

A Chu Hoi pass.

Viet Cong and North Vietnamese Army troops could present this document and receive safe passage to Allied lines.

The remnants of the 7th Artillery that were not transferred or discharged, took the unit's regimental colors, and little else back to Fort Riley, Kansas. Among those accompanying the colors was Sergeant Bob Abbott. The unit continued as the 7th Artillery until August 13, 1971 when a message was received at battalion headquarters from the commanding officer of the Institute of Heraldry, U.S. Army. It reads:

Subject: Reorganization of CARS (Combat Arms Regimental System) Artillery Regiments.

1. Based upon [a] plan developed by Chief of Military History, the Chief of Staff, U.S. Army has approved [the] reorganization [of] existing CARS Artillery regiments into CARS FIELD ARTILLERY and AIR DEFENSE ARTILLERY regiments. Effective 1 Sep 71, repeat 1 Sep 71.

2. On 1 Sep 71, 7th Arty will become 7th FA (field artillery) Regt and 7th ADA (air defense artillery) Regt respectively. Battalions of present 7th Arty with assigned FA mission and equipment will be organized as corresponding battalions, 7th FA and battalions assigned with AD (air defense) mission and equipment will be organized as corresponding battalions, 7th ADA Regt. Accordingly, 1st Bn, 7th Arty will become 1st Bn, 7th FA Regt, 3d Bn, 7th Arty will become 3d Bn, 7th ADA Regt; 8th Bn, 7th Arty will become 8th Bn, 7th ADA Regt. Heraldic entitlements existing prior to organization of 7th Arty on 20 Dec 1965 will be restored eff 1 Sep 71.[56] 7th FA Regt is lineal descendant of 7th FA organized 3 June 1916, and 7th ADA Regt is lineal descendant of [the] 7th Coast Artillery originally organized 8 March 1898.[57] Coat of arms and distinctive unit insignia for these two regiments which were heraldically cancelled 20 April 1960, are restored and officially authorized effective 1 Sep 71.

3. In view of above CO, USASPTC, Philadelphia, will prepare job order for new organization color and color will be forwarded directly to you when completed. Present organization color will be used until receipt of new color at which time

---

[56] This sentence delivered the death blow to the Pheons as the heraldry that existed before Dec 20, 1965 had no pheons in it.
[57] This sentence effectively made the 7th Artillery and its distinctive unit insignia an orphan, with no parent unit.

present color will no longer be used and will be disposed of IAW (in accordance with) per (paragraph) 116, AR (army regulation) 840-10.
4. Distinctive unit insignia authorized for 7th FA on 26 Feb 1923 will be used by 1st Bn, present 7th Arty, and that auth for 7th CA on 31 Jul 1924 will be used by 3d and 8th Bns, present 7th Arty. It is personal desire CSA[58] that historic heraldic entitlements of 7th FA and 7th CA be utilized by [the] new regiments.

So, with two separate types of artillery regiments, one field artillery and one air defense artillery, which would inherit the Vietnam War campaigns of the 7th Artillery? The surprising answer is both. And the 7th Artillery? Well it was sort of erased. It's not a parent unit of either of the units as their creation predates the creation of the 7th Artillery.

The Army's institute of heraldry just wrote my regiment out of its own history and awarded its honors to the other two units. It reminds me of the Egyptians when they would chisel the name of a despised pharaoh from temple walls. The only mention of the 7th Artillery is in the listing of Army mottoes. The motto "Nunquam Fractum" is there, with OBS written beside it inside parenthesis. When I wrote to the institute asking them what that meant, they said it is an acronym for "obsolete". They explained it was because the unit insignia had been canceled, as if it had never been.

This kind of thing is best summed up by an anonymous poem written on the wall of a guard shack many, many, years ago. It goes like this.

---

[58] Chief of Staff, U.S. Army.

God and the Soldier we adore,
In times of Trouble and no more,
When the danger is past and all things righted,
God is forgotten and the Soldier is slighted.

7th Artillery Battery Commanders in 1969

Headquarters Battery

1 Jan 69 – 15 Mar 69	CPT Kenneth J. Szuszka
16 Mar 69 – 16 Sep 69	CPT Michael A. Huston
16 Sep 69 – 31 Dec 69	CPT William J. Forbes

Alpha Battery

1 Jan 69 – 10 Mar 69	CPT John E. Dubia
11 Mar 69 – 1 Sep 69	CPT Robert S. Mager
1 Sep 69 – 31 Dec 69	CPT Harry G. Madden

Bravo Battery

1 Jan 69 – 9 Apr 69	CPT Wayne A. Miller
10 Apr 69 – 19 Sep 69	CPT Larry R. James
20 Sep 69 – 31 Dec 69	CPT George Kellendenz

Charlie Battery

1 Jan 69 – 10 Jan 69	CPT William McCain
11 Jan 69 – 16 Jul 69	CPT John Todd
17 Jul 69 – 31 Dec 69	CPT James M. Kelly

Vietnam, the War That Never Ends

After returning from the war I remained in the U.S. Army for another twenty-seven years including the Regular Army and as an Active Guard/Reserve (AGR) member. During that time I served overseas in Europe, completed a tour as an Army recruiter, and returned to Oklahoma as an active duty Army administrator for a Stillwater, Oklahoma based infantry brigade of the 95th Infantry Division (USAR).

With a year's break to attend the U.S. Army Sergeant's Major Academy[59], I worked with the 3rd Brigade of the 95th Division in Stillwater from October 1979 until my retirement and transfer to the Retired Reserve for an additional nine years and subject to recall, at the end of June in 1988. Looking at a promotion to sergeant major, the top enlisted rank, with a reassignment to a large city, I opted to remain in Stillwater and retired at the rank of master sergeant.

At the time of retirement I was serving as the senior infantry operations sergeant for the brigade. During that eighteen years I was awarded five Good Conduct Medals, two Army Achievement Medals, three Army Commendation Medals, and the Meritorious Service Medal in addition to the two Bronze Stars, Vietnam Service Medal with 4 campaign stars, Vietnam Campaign Medal and the Vietnamese Cross of Gallantry with Palm (awarded to the 7th Artillery and all members of the unit), and the Vietnam Civil Actions Medal, earned in Vietnam. Throughout that time frame I fired expert with the M-16A1 rifle every year. In May of 1997 I received my final Honorable Discharge from the United States Army.

The last words in the documentary *Inside the Vietnam War* sums up the feelings of many Vietnam veterans:

---

[59] Only the top 4 percent of the Non-commissioned officers in all the Armed Forces is selected to attend the Sergeant's Major Academy.

The Vietnam War requires respect for its hopes, understanding of its heartaches, and honor for its heroes.

I had served my country well and with honor for more than twenty years, and now I looked forward to finding me and my family's place in the American dream. Unfortunately, I chose to work for the wrong employer, Oklahoma State University. They would never let the Vietnam War end for me. I received an honorable discharge on May 14, 1997, after almost thirty years of service.

The Vietnam Service Medal

These two photographs are from Fire Support Base Huertgen near the Iron Triangle. Redleg's rarely wore shirts. I was all of 135 pounds soaking wet! The writing on the photo is my dad's. Bandido Charlie Company 1st Battalion 16th Infantry was with us here.

This is a good illustration of the older type of howitzer (M-101) with its sandbag parapet surrounding the position. The flag in the background is the Oklahoma state flag.

Captain John Dubia and his XO, Battery A commander, FSB Huertgen. The battery's guidon showing its designation as A, 1/7th is clearly seen.

Not readily visible in this photo is the fact I had a "new guy" sunburn. This is the first photograph of me in the field. It's Miller time!

I kept my driving skills sharp by driving the battery commander's jeep inside FSB Aachen. We were now in the Iron Triangle.

I and a fellow redleg stand inside a gun position termed a "parapet" at FSB Aachen. The guard tower is clearly visible in the background. Note the tip of artillery rounds or "projos", short for projectiles, behind us. This photo was taken in March, 1969. This is a well constructed gun position.

Artillerymen are nicknamed "Redlegs" a name given them by the infantry because of the uniform they wore during the American Revolution that consisted of a blue coat and red trousers.

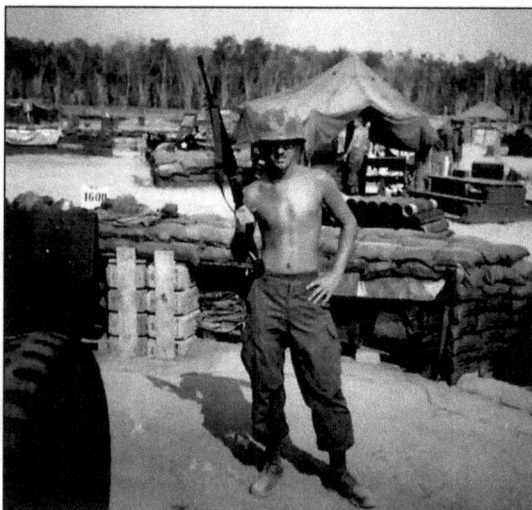

Every soldier I ever knew has done one of these "Machine Gun Kelley" kinds of photos. The tent right behind me was my home. You can see the jeep I was in for the earlier picture over my right shoulder. The tree line behind me was about 300 meters away, which gives you an idea of the size of the rubber trees surrounding Fire Support Base Aachen in the Iron Triangle.

Another photograph similar in nature at FSB Aachen, March 1969.

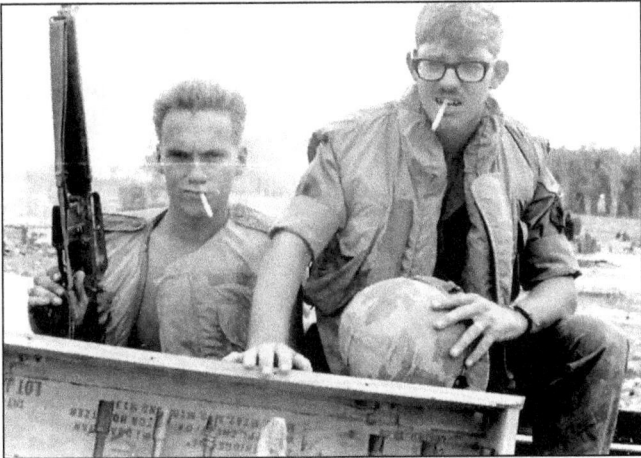

The note on the back of this photo, taken and sent home to my parents, says this soldier is my best friend. I've forgotten his name but in my later narrative he is referred to as "the killer". We are moving by convoy to Fire Support Base Jim, April 1969. My hand rests on a box containing loaded M-16s.

Scenes from Fire Support Base Jim: left a fire mission is in progress, note the dust rising from the concussion of the gun's firing and the gun in recoil. On the right is our well constructed shower. Water, heated by the Vietnam sun in 10 gallon water cans, was poured into the shower bucket. This excellent shower would fall prey to an enemy 82mm mortar.

An early scene at FSB Jim taken in May of 1969 outside the Exec Post, the battery guidon flutters above my head. This kind of head gear was called a "go to hell" hat by the soldiers but was actually termed a "slouch" hat.

In this bull session, the sergeant standing in the middle of the photo, like myself, was recommended for the award of the Silver Star for "Gallantry in Action". He also earned the Purple Heart. The oldest man in this photo is 21 years of age. The kid sitting down was a very cocky individual. But then I guess we all were.

Ready for action are artillerymen John Jordan, Thomas Bonk, and Paul Jones. Paul and I correspond to this day and he still refers to me as "Doc".

Doc's office: that is a patient in my "waiting room" with my medical supplies behind him and my M16 above him on the wall.

August 31, 1969 at Fire Support Base Jim. The tent behind me was my "hooch". I'm wearing unauthorized airborne combat boots that zip up on the inside. Note the friendship bracelet given to me by Babysan, the daughter of our laundress on my right wrist. It was not long after this photograph that the battle for FSB Jim took place.

The water puddle behind us indicates it just rained. First Sergeant Jerry Cooper sports a band aid from a small wound received during the fight for FSB Jim, September 1969. The supply sergeant to my left brought us steaks and beers from the base camp at Phuoc Vinh. Both men received BSMs.

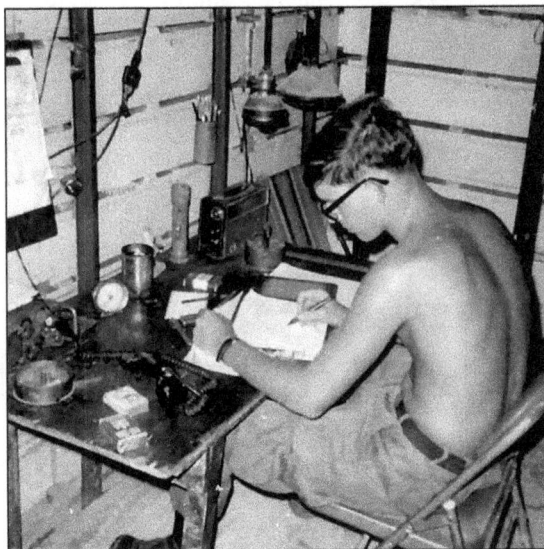

Here I am in the Exec Post recording a fire mission. The ashtray was made from the base of an expended artillery round. I have an engraved one to this day given to me by Major Bob Church S-1 officer of the battalion (1965 – 1966). The radio handset lies on the field desk.

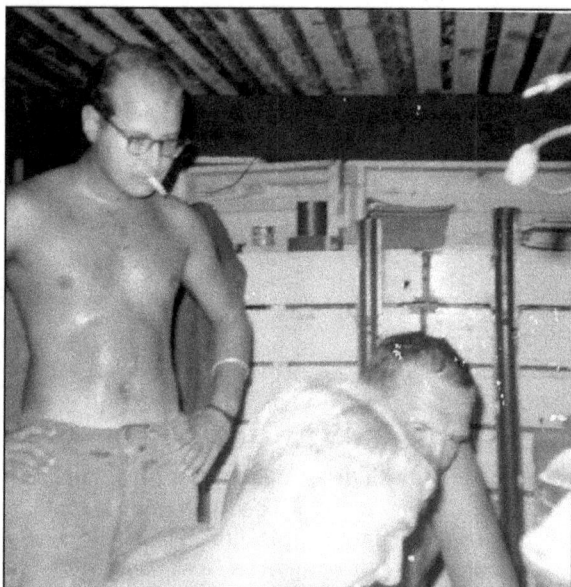

Exec Post members Poteet standing and Nagley sitting with First Sergeant Cooper. Poteet was known for his seemingly endless grumpiness. Nagley was always smiling as he is here.

Another shot of the shower stall about to be put to good use.

In this photograph I'm decorated by Colonel Sperow at Fire Support Base Jim, it was an exceptionally beautiful and cool day in Vietnam. I had just been promoted to Specialist 5 (same pay grade as a sergeant) November '69.

My bunk inside the "hooch" I shared with two other soldiers.

Exec Post clerk Nagley is impersonating the battery commander, Captain Harry G. Madden

The real Alpha Battery Commander, Captain Harry G. Madden

The average age of the men of Alpha battery was twenty years old. Captain Madden wasn't much older, perhaps his late twenties, but despite his age, he was the "old man". It was Captain Madden, with concurrence of the battery first sergeant that recommended me for the Silver Star Medal. We sometimes called him the "BC".

Behind the captain you can see a stream of water from the monsoon runoff. A heavy gauge steel construction piece known as PSP forms a foot bridge for easy passage. PSP was used throughout Vietnam from aircraft runways to the floors and roofs of bunkers. It was one of the most commonly used types of construction material besides steel engineer stakes and sandbags.

A hot day somewhere in Vietnam November 1969. Note the friendship bracelet is gone. I mailed it to Carol for her little sister Rhonda.

Last photo taken at Fire Support Base Jim, in front of my medic trailer. Soon after this picture we headed out to Fire Support Base Florida, December '69.

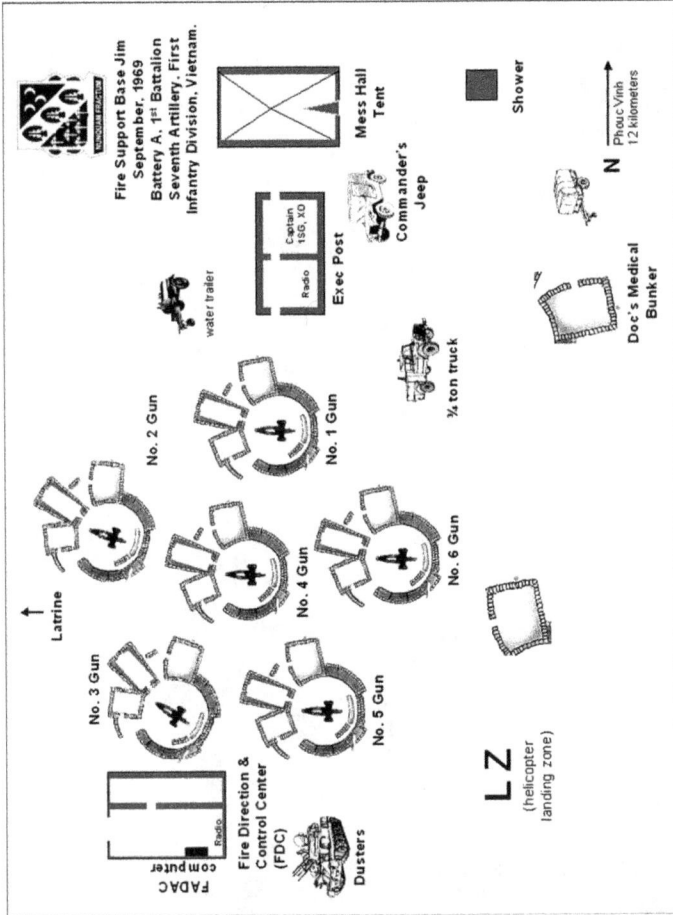

Diagram 1, Fire Support Base Jim September 8, 1969.

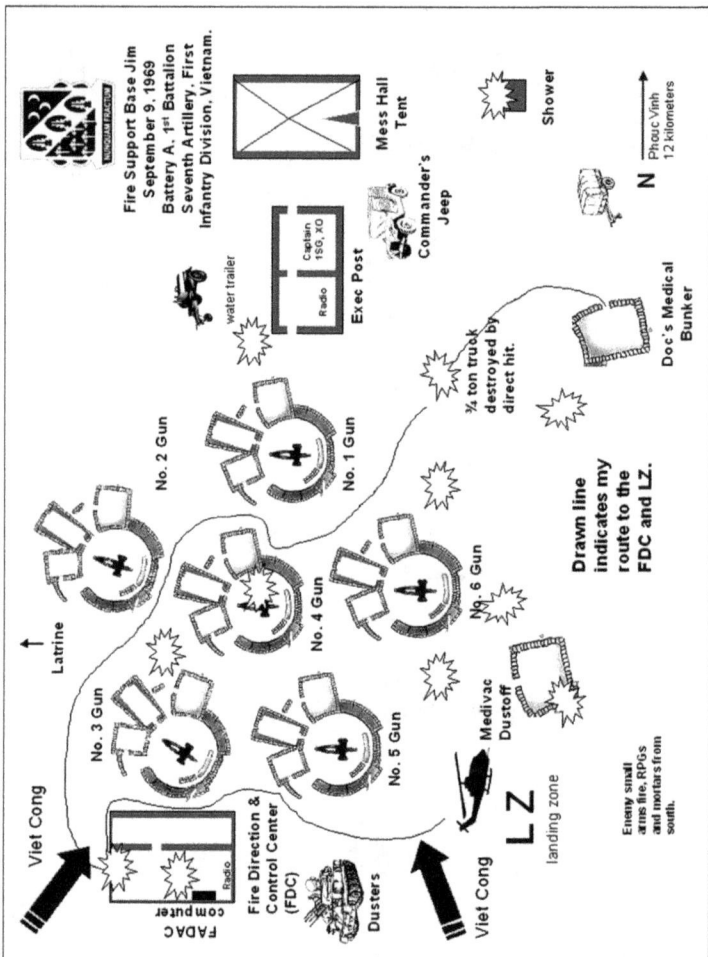

Diagram 2, Fire Support Base Jim September 9, 1969.

Base plate of a 105mm round fired by the 7th Artillery early in the war, sometime in 1965 or 1966 courtesy of Bob Church. Note the date of "1945" indicating this round was created while my dad, a WWII veteran, was still in the Army. This "trench art" ashtray sat on Bob's field desk in Vietnam.

Some of the engraving on the ash tray given to me by Bob Church. Also on the sides is the unit DUI and "1st Battalion 7th

Artillery" all filled in with artillery red. It is one of the prized possessions from my military career.

In February of 1968 Specialist Russell Gallegos arrived in country and was assigned as an artilleryman to Battery A, 1st Battalion, 7th Artillery. He recalls his arrival in his own words.

"When we arrived in Vietnam (aboard a commercial airliner) we could not land because the airfield was under mortar attack. We detoured to the Bien Hoa Air Base where we would have ended up anyway.

There, overnight, we got in a deuce and a half (2 and ½ ton truck) headed for Di An. Upon arrival I was issued an M-14, web gear, flak vest, and assigned to a gun crew. I changed into jungle fatigues and put my personal effects in a locker. Staff Sergeant Growler, my new crew chief, collected me and my first detail was to fuse a load of 105mm rounds and store them in a bunker. There was a blonde headed corporal I later learned was from Ohio named Edward Koehn, he eye balled me and said 'Get your ass in here and get to work Poncho Villa' the nickname stuck, Ed Koehn and I remain friends and close as brothers to this day. I'd do anything for him, and although he lives 1500 miles from me, we keep in touch."

The battery supported elements of the 1st Infantry Division who secured the massive Tan Son Nhut Air Base during the communist offensive of Tet 1968. Each day of the offensive the battery's round output doubled from the day before.

The following month Major General Keith L. Ware too command of the division as the Big Red One took part in Operation Quyet Thang (Resolve to Win). One month later and the division was involved in the largest operation of the Vietnam War, Operation Toan Thang (Certain Victory). A scant six months after taking command, on 13 September 1968 General Ware was killed in action when his command helicopter was shot down by enemy anti-aircraft fire. Battery A, and other elements of the battalion fired a continuous fire mission until the general's remains and those of his crew and staff were recovered.

The division assistant commander, Major General Orwin C. Talbott assumed command of the division.

In the meantime Russell found himself outside the village of Cu Chi supporting the 25th Infantry Division and serving as the radio operator for a forward observer. In an old Buddhist temple he met a medic named Joaquin, also from Texas. They took a liking to one another and together sang Mexican songs in the evening. Joaquin made fun of Russell because he could not speak Mexican Spanish.

While on patrol Russell was shocked to discover a dead ARVN soldier hanging by his neck. No one knew if the VC had done it or maybe other Vietnamese soldiers. They also found an enemy weapons cache and Russ helped himself to an AK-47 that he carried throughout the remainder of his duty tour.

When Russ returned to the battery's firing position he helped fire a mission in support of the very outfit he had been on patrol with. During a close combat firefight Joaquin was hit and killed in action. Russ was saddened by the death of his new friend. In his honor Russ later named his son Christopher Joaquin Gallegos.

The battery rotated back to Di An and then on to a new position in the Iron Triangle. But not before enemy sappers came on the base throwing satchel charges in the barracks. The battery fired illumination and set up to use killer junior, but the enemy just melted away.

Russ found the assistant gunner had rotated home and he now took that position. The gun chief told Russ he wanted their gun to be the fastest gun in the battalion. He had Russ work become closely familiar with the hook, rope, and wooden handle of the artillery gun lanyard. Once he had its operation down the gun was firing between fifteen and eighteen shells per minute. That was fast!

During the TET offensive the battery was constantly moving to support elements of the 25th Infantry Division, the 1st Cavalry Division and our own Big Red Unit units. Sometimes they would move forward in support of the infantry and then back to the base in the same day. When the battery moved the mess hall went with them, and Russ states they only ate combat rations once.

During our long stay at Fire Support Base Jim the mess hall did come out for awhile. The head cook, a sergeant first class, worked at night doing the baking during the cool hours. Part of my job as the medic was to insure the sanitation of the mess hall. So sometime during the night, usually when I could smell the pastries baking, I inspected the mess tent. And of course I had to sample those pies that were cooking and make sure they were hot and delicious enough, especially those cherry pies! They always were.

Russ and the battery pulled into a new position near the Cambodian border and in late June of '68. The first order of business was to build a gun parapet. It took three days and nights to fill the fifteen hundred sandbags to make a good fighting position, doing the work between hot fire missions. And just as the parapet was completed the battery was moved to An Loc.

It wasn't long before an emergency fire mission came down. An infantry company was in danger of being overrun by a strong Viet Cong force. The infantry called for fire on their own position and although we received no official word on casualties either friendly or enemy we got a thank you for getting the VC off their ass. It was hard to sleep for the next few nights.

"Back in the Iron Triangle and near the Black Virgin Mountain (Nui Ba Den) we moved in support of the Rome plows. The battery was airlifted into positions to protect the engineers operating the plows which were stripping the jungle away. The VC was not happy about being exposed to our aircraft and ground forces. With the removal of trees and brush the spiders, centipedes, scorpions and snakes were looking for new homes, with us GIs. We followed the plows for two weeks until we came to a road. There the battery loaded up on trucks for a road movement. While moving ammunition I and two other battery members were stung by scorpions. The medics treated us with aspirin, water, and rest and we climbed up inside the truck and went to sleep, a little unexpected in country R&R.

During the monsoons I managed to get a bunk space in a tent and moved out of the rat infested bunker I had been staying in. About eight o'clock that evening I heard a distinct thud and I knew the rain had caved a bunker in. I rushed and found it was the bunker Marvin Schwint was sleeping in. I started pulling sand bags off not sure which end of the bunker he was in, and I called for more help. I had got lucky and guessed correctly and after removing only five or six sand bags I found him. He gasped for air as I pulled him up and I noted a wound on his forehead where the steel reinforcement of the ceiling had hit him. He was air lifted to a field hospital and was okay. When he came back he told me I was the prettiest thing he had ever seen when I pulled that sand bag off him."

Sergeant Russ (Poncho) Gallegos
Battery A, 1st Battalion, 7th Artillery (Pheons)

"I am glad to have served with some real American soldiers."

Specialist (later sergeant) Russell Gallegos and Sergeant Jim Frisk at Lai Khe, Russell provided the following important photos after contacting me in September 2012. Our tours of duty only overlapped by a month (Jan – Feb 1969), I am grateful to him for providing them.

In a ceremony in September 1968 Major General Albert E. Milloy came to Battery A, 1st Battalion, 7th U.S. Artillery in Lai Khe to fire the one millionth round sent down range by the battalion. Preparing for his arrival is a four man gun crew. The first two soldiers from the left are Sergeant Jim Frisk and another soldier identified only as Moon. Sergeant Frisk's gun was selected to fire the millionth round because his gun was the oldest howitzer in the 1st Infantry Division.

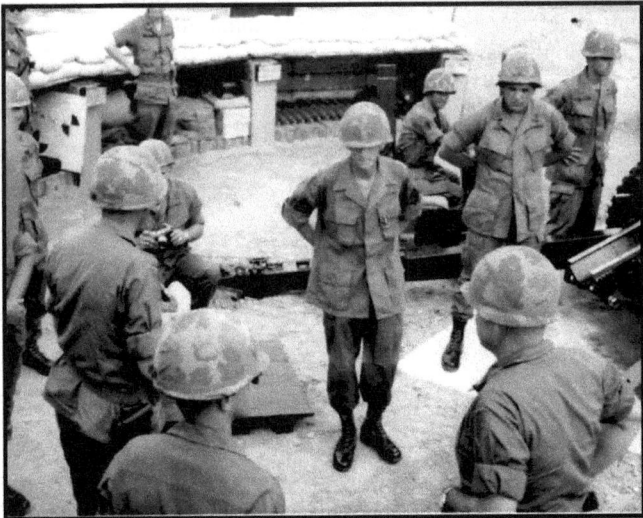

General Milloy (far right back to camera) arrives for the ceremony accompanied by battalion commander Lieutenant Colonel Merle M. Crocker (first full figure from left back to camera). The man just to the left of the general is likely the battery commander. The soldiers from left to right (center) are Sergeant Jim Frisk, Moon, Robert Chappa and an unknown soldier.

General Milloy pulls the lanyard and fires the one millionth 1st Battalion, 7th Artillery round to be fired in Vietnam. There would be another quarter million rounds fired before the battalion left for the U.S. The Oklahoma and California state flags fly in the background.

Left to right: Staff Sergeant Growler, Bob Reed, Russ Gallegos, Ron Mallard, Bob Pollack, Ed Koehn, unknown. February 1969 in Di An. Russ remembers names better than I.

Russ and other gun bunnies distribute artillery shells to the separate gun sections. This appears to be an Iron Triangle position, likely FSB Huertgen.

Russ' caption for this picture reads "Ed Koehn and Doc, Iron Triangle". No, that's not me. This is the medic I replaced in January 1969 when he was either wounded or injured. Everyone knew him (and later me) as "Doc" and no veteran of the 1/7th I've contacted remembers his actual name.

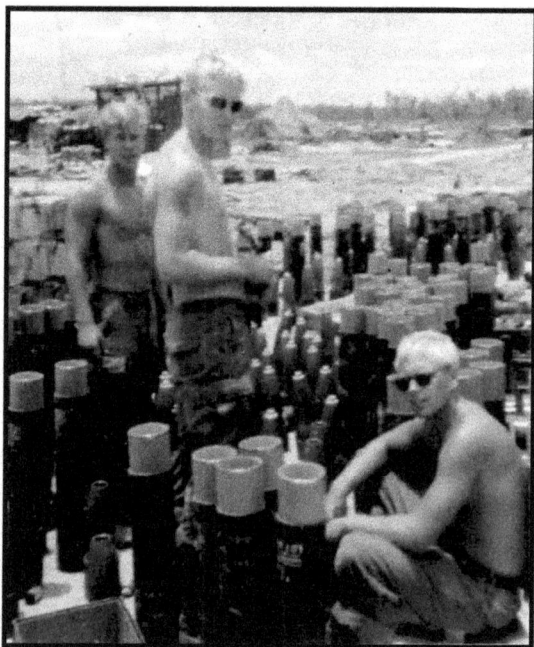

Left to right is Corporal Ed Koehn, unknown, and Ron Mallard
assembling projectiles at FSB Huertgen, January 1969.

Following the Rome plows in the Iron Triangle near Nui Ba Den
(Black Virgin Mountain, 996 meters high of solid rock with a
thin veneer of soil), the battery was airlifted in to support jungle
clearing operations. In this photograph a Chinook helicopter
brings in one of the battery's 105mm howitzer, its ammunition
conveniently slung underneath. This mission lasted two weeks.
Russ and others got stung by scorpions here.

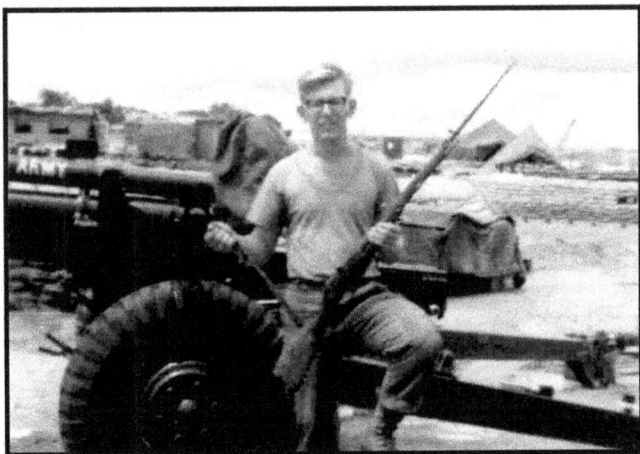

Bob Pollack poses with an enemy rifle captured during the Rome
plow operation. It is not the famous AK-47 which is an
automatic rifle. It is a bolt action rifle, possibly a rifle left over

from the Second World War. In some instances these kinds of weapons could be taken home as souvenirs, but full automatic weapons like the AK were forbidden.

Fire mission! This looks like a Thunder Road position, that's Russ with his face an inch or two from the sight, Bob Mallard standing with his back to the camera, Bob Reed in the background. The other two men are unknown. A spent round lies on the ground with others on the ready board awaiting the designated powder charge (1 to 7) and type of fuse, usually, fuse quick. Note the sand bags atop the ammo bunkers.

End of fire mission. Left to right unknown, Bob Reed, Russ, Bob Mallard, unknown.

Joey Bishop visits the battery in November of 1968. Russ says he was a nice guy, humbled that he was among fighting soldiers. Bob Pollack (glasses, looking away from camera, Bob Mallard in background, Staff Sergeant Growler, Joey Bishop with Russ standing directly across from him (back to camera), Staff Sergeant Downing, and Corporal Ed Koehn. I never heard this visit by Joey Bishop mentioned, and I arrived only about a month later.

Iron Triangle, January 1969. That appears to be a command Loach helicopter flying into the battery area. It's quite likely the battalion commander (by that time Lieutenant Colonel Francis King) visiting the battery which had just come under the command of Captain John Dubia. And around the time the battery received a new "Doc".

Sometimes the engineers were slow removing trees that could provide a hiding place for the enemy. A quick "direct fire" mission could remove a tree quite handily, and provide a little target practice for the gunners.

The last picture Russ took in Vietnam of the 1/7th Artillery.

I am a member of the Society of the First Infantry Division and as such I receive quarterly issues of the society's magazine the *Bridgehead Sentinel.* In the fall 2014 issue there appeared a recurring section titled *Bro Books* that invited the submission of books written about units of the Big Red One. I submitted the then current edition of this book and it appeared in the winter 2014 issue. A number of 7th Artillery veterans ordered the book, and with each copy sent I invited them to submit photographs and or narratives they would like to have in a follow up edition. What follows here are those photographs and stories submitted to me for publication. I want to thank those who have contributed to this living document.

Corporal Don Burwell is one of the men who read the book and responded with photographs and narrative. Don was one of the original members of the 1/7th Artillery to arrive in Vietnam serving from November 1965 to November 1966. He flew out of Tay Ninh near the Black Virgin Mountain taking part in *Operation Attleboro.*

"Yes, I'm an early pioneer of the Big Red One in Vietnam. We did two amphibious landings on the beaches of the South China Sea near the mouth of the Saigon River and [the village] of Rung Sat. It was like a miniature D-Day landing. [It was] kinda cool doing it.

Our main mission was to provide fire support to the engineers while they bulldozed base camp areas. One [was at] Bearcat, [and] also while they were expanding Di An. We never got to live in any permanent buildings, just on the ground or sometimes on cots because we moved on before the camps were finished...It was actually cleaner out in the field on operations than it was at the very muddy bulldozed base camp areas. We did many air assaults with our gun[s] hanging beneath a Chinook helicopter or inside a C-123.

C-123 *Provider*

Don Burwell aboard 1/7th landing Craft – South China Sea

Don was involved in two amphibious landings on the beach from the South China Sea and the Saigon River when the 1/7th first arrived in Vietnam.

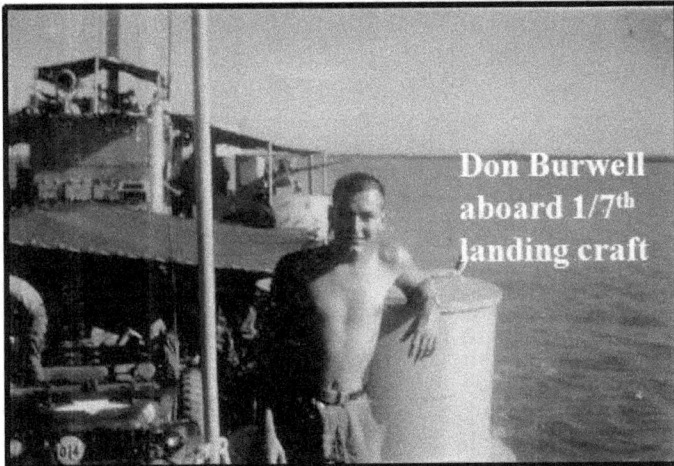

Don Burwell aboard 1/7th landing craft

Coming up the Saigon River and almost in Vietnam.

Corporal Don Burwell Battery A, 1/7th Artillery in 1965 carrying a Thompson .45 caliber submachine gun.

When I saw this picture it reminded me of the incident of our supply sergeant offering me a Thompson submachine gun in 1969. So I asked Don about it and he replied.

"Yes that is a Thompson in the photo of me... I was issued an M-14 rifle, but it [the Thompson] was given to me by my buddy, Corporal James Bates when he went back to the states, and then it was turned in to supply...[1] Since it was the only Thompson in our Battery, it was probably the same gun... Small world..."

---

[1] In 1966 when Don Left, and then offered to me in January of 1969.

Don Burwell here posing with an A Battery M-101 howitzer in 1966. Note that the gun appears to be in an open field that without a fortified perimeter.

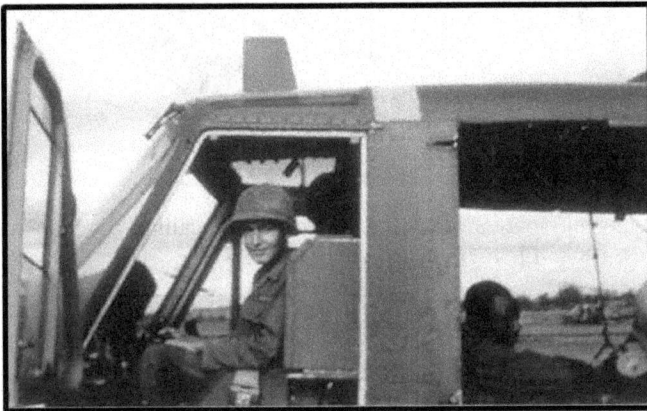

Don Burwell in a Huey helicopter, ready to head home.

After his service in Vietnam Don returned to the U.S. and a new assignment Fort Carson, Colorado. There he trained and served as a gunner on the M-109 self propelled howitzer until his honorable discharge from the U.S. Army.

Lieutenant John Swartz served in the 7th Artillery from 1967 to 1968 as a forward observer with the infantry. Captain Cary S. King served as the battery commander of Battery C in 1968. Both these men were with the 1/7th during the intense action of 'TET 1968'. These are the photographs they sent me for inclusion in this history.

Lieutenant John Swartz, forward observer, calling for blocking fires, 1968. Photographs courtesy John Swartz.

Operations and Intelligence section at Di An in 1967.

Battalion Headquarters at Di An looking east in 1967. Photograph courtesy John Swartz.

C Battery at Fire Support Base Sicily VI. Photograph courtesy John Swartz.

C Battery in action during TET '68, location unknown. Photograph courtesy Captain Cary S. King.

Battery C during TET '68. Photograph courtesy Captain Cary S. King.

This is my friend Tom Bullock who served as a member of the 377th Security Police Squadron at an air base near Saigon in 1969. Tom was in the Air Force at that time. While at FSB Florida, I was only about 20 kilometers from his position. When the base came under enemy probing or attack, we [Alpha, 1/7] provided supporting fire for Tom's unit and the rest of the air base. It would be nearly forty years before I would meet Tom and exchange war experiences discovering this link between us. I met Tom when I went to Washington D.C. for ESGR training in 2007. Photograph courtesy Tom Bullock.

At the Vietnam Veterans Memorial Washington, D.C. September 2008. The day I visited it was very hot and humid, appropriate Vietnam weather. In May of 2010 six more names were added to the wall. These were veterans of the war who sustained wounds so severe that they eventually succumbed to them years later. They are Army Lt. Col. William Taylor, Marine Corps Lance Corporals John Granville and Clayton Hough Jr., Marine Corps Corporal Ronald Vivona, Army Captain Edward Miles and Army Sergeant Michael Morehouse, comrades all.

The artillery adds a bit of class to what would otherwise be a vulgar brawl. This is the Distinguished Unit Insignia (DUI) of the 7th Artillery. The motto "Nunquam Fractum" is Latin for "Never Broken". The three arrowheads are called "Pheons" in heraldic terms and that was the nickname of the men who served with the 7th Artillery in Vietnam. It was a name that would remain in Vietnam, because when the unit returned stateside in March of 1970 it was redesignated the "7th Field Artillery" and the new insignia, assigned by the U.S. Army Institute of Heraldry, did not include the pheons.

The 7th Artillery pocket patch. Normally, unit insignia is made of metal and worn on the shoulder strap of the dress uniform. In Vietnam it was cloth and worn on the left pocket of the combat fatigue uniform shirt. The colors are silver and red. Red is the official color of the United States Army Arillery. The Army hymn was initially the official song of the artillery. St. Barbara is the patron saint of the artillery.

Master Sergeant Walter L. Cross

Official photograph, U.S. Army Sergeants Major Academy. My promotion packet for sergeant major went forward in February 1988. However, the family decided it would be better for us all if I retired rather than be transferred away from our Stillwater home.

Appendix A: Reorganization of 7th U.S. Artillery Regiment

JOINT MESSAG' RM						SECURITY CLASSIFICATION UNCLASSIFIED		0614212 *Aug*

PAGE	DRAFTER OR RELEASER TIME	PRECEDENCE		LMF	CLASS	CIC	FOR MESSAGE CENTER/COMMUNICATIONS CENTER ONLY	
		ACT	INFO				DATE – TIME	MONTH YR
1 OF 4	051200Z	RR	RR				*1452 Aug 71*	

BOOK Yes		MESSAGE HANDLING INSTRUCTIONS

FROM: CO THE INSTITUTE OF HERALDRY US ARMY CAMERON STA VA

TO: CO 1ST BN 7TH ARTY FT RILEY KS

CO 3D BN 7TH ARTY SCHWEINFURT GERMANY

CO 8TH BN 7TH ARTY FT BLISS TX

INFO: CG 1ST INF DIV FT RILEY KS

CGUSARFIVE FT SAM HOUSTON TX

CINCUSAREUR

CG FT BLISS TX

UNCLAS AGAN-C. Subj: Reorganization of CARS Artillery Regiments.

THIS MESSAGE IN TWO PARTS.

PART ONE FOR ALL.

1. Based upon plan developed by Chief of Military History, the Chief of Staff, US Army has approved reorganization existing CARS Artillery regiments into CARS FIELD ARTILLERY and AIR DEFENSE ARTILLERY regiments. Separate instructions will be issued by DA relative to reorganization of existing regiments, effective 1 Sep 71, repeat 1 Sep 71. Accordingly, this msg relates only to heraldic entitlements for separate regiments concerned.

DISTR:

13 AUG 1971

DRAFTER TYPED NAMED, TITLE, OFFICE SYMBOL AND PHONE	SPECIAL INSTRUCTIONS
EUGENE F. GANLEY COL AGC AGAN-C 274-6630	
TYPED NAME, TITLE, OFFICE SYMBOL AND PHONE	
EUGENE F. GANLEY COL AGC AGAN-C 274-6630	
SIGNATURE	
	SECURITY CLASSIFICATION

PAGE	DRAFTER OR RELEASER TIME	PRECEDENCE		LMF	CLASS	CIC	FOR MESSAGE CENTER/COMMUNICATIONS CENTER ONLY		
		ACT	INFO				DATE – TIME		MONTH	YR
2 OF 4	051200Z	RR	RR						

JOINT MESSAGE **FORM**

SECURITY CLASSIFICATION
UNCLASSIFIED

BOOK
Yes

MESSAGE HANDLING INSTRUCTIONS

FROM:

TO:

2. On 1 Sep 71, 7th Arty will become 7th FA Regt and 7th ADA Regt respectively. Battalions of present 7th Arty with assigned FA mission and equipment will be organized as corresponding battalions, 7th FA and battalions assigned with AD mission and equipment will be organized as corresponding battalions, 7th ADA Regt. Accordingly, 1st Bn, 7th Arty will become 1st Bn, 7th FA Regt, 3d Bn, 7th Arty will become 3d Bn, 7th ADA Regt; 8th Bn, 7th Arty will become 8th Bn, 7th ADA Regt. Heraldic entitlements existing prior to organization of 7th Arty on 20 Dec 1965 will be restored eff 1 Sep 71. 7th FA Regt is lineal descendant of 7th FA organized 3 June 1916, and 7th ADA Regt is lineal descendant of 7th Coast Artillery originally organized 8 Mar 1898. Coat of arms and distinctive unit insignia (para 14-19a, AR 670-5) for these two regiments which were heraldically cancelled 20 Apr 1960, are restored and officially auth eff 1 Sep 71.

3. In view above CO, USASPTC, Philadelphia, will prepare job order for new organizational color and color will be forwarded directly to you when completed. Present organization color will be used until

DISTR:

DRAFTER TYPED NAMED, TITLE, OFFICE SYMBOL AND PHONE

SPECIAL INSTRUCTIONS

TYPED NAME, TITLE, OFFICE SYMBOL AND PHONE

SIGNATURE

PAGE	DRAFTER OR RELEASER TIME	PRECEDENCE		LMF	CLASS	CIC	FOR MESSAGE CENTER/COMMUNICATIONS CENTER ONLY		
		ACT	INFO				DATE – TIME		MONTH YR
3 OF 4	051200Z	RR	RR						

JOINT MESSAGE FORM

SECURITY CLASSIFICATION
UNCLASSIFIED

BOOK

MESSAGE HANDLING INSTRUCTIONS

FROM:

TO:

receipt of new color at which time present color will no longer
be used and will be disposed of IAW par 116, AR 840-10.

4. Govt owned dies to be used in the manufacture of your regimental
distinctive unit insignia are available for use by certified manu-
facturers. You should take immediate action to commence procurement
of new insignia from NAF.

5. Distinctive unit insignia authorized for 7th FA on 26 Feb 1923
will be used by 1st Bn, present 7th Arty, and that auth for 7th CA on
31 Jul 1924 will be used by 3d and 8th Bns, present 7th Arty. It is
personal desire CSA that historic heraldic entitlements of 7th FA and
7th CA be utilized by new regiments. Accordingly, no change coat of
arms and distinctive unit insignia of old 7th FA and 7th CA auth.

6. Separate ltr of instructions on procurement of distinctive unit
insignia follows. Auth ltr for subj insignia eff 1 Sep 71 follows.
Grant of arms with accompanying organizational painting will be issued
subsequent to 1 Sep 71.

DISTR:

DRAFTER TYPED NAMED, TITLE, OFFICE SYMBOL AND PHONE

SPECIAL INSTRUCTIONS

TYPED NAME, TITLE, OFFICE SYMBOL AND PHONE

SIGNATURE

SECURITY CLASSIFICATION

7th Field Artillery

7th Air Defense Artillery Battalion

Appendix B: Journal Entries; Headquarters, 1/7th Artillery Dian, Vietnam for September 9, 1969.

DAILY STAFF JOURNAL OR DUTY OFFICER'S LOG

CONFIDENTIAL

ORGANIZATION OR INSTALLATION	LOCATION	PAGE NO.	NO. OF PAGES
S-3 Section 1st Bn, 7th Arty	Di An, Vietnam XT9007	2	17

PERIOD COVERED			
FROM: HOUR	FROM: DATE	TO: HOUR	TO: DATE
0001	9 Sept 69	2400	9 Sept 69

ITEM NO.	TIME IN	TIME OUT	INCIDENTS, MESSAGES, ORDERS, ETC.	ACTION TAKEN	INITIALS
(4 cont)	0143		A 1/7 says FDC blown away & probably could not fire a mission at this time		
	0148		Dustoff to arrive from NW in 10 min		
	0151		Radisa C 8/6 from N-S 1000 meters to East to be adjusted from 889386		
	0152		A.O. 50 E Due on station 10 m.		
	0155		Req. E.O.M. on HE from C 8/6 incoming Stopped Hunter-Killer to work around perimeter EXP 16 NEPO		
	0155		30 wants arty to the east of Hgwy through FSB Jim - Hunter-Killer to work western ½		
	0158		Dynamite A 8/6 DNL 884387 up & ready		
	0200		NIGHTHAWK in area will go take a look		
	0210		Sit-Rep. Sapper got through wire & hit FDC		
	0217		A.O. 50 E out of Phu Loi		
	0218		A 1/7 to cut-off Killer Jr. for Hunter-Killer team and Spooky to work over area EXP 368 HE/MT 140 SELF ILLUMINATION		
	0250		Found 5 ChiCom grenades outside gate " 3 on FDC Bunker Some thrown inside FDC - about 9 grenades all around bunker		

TYPED NAME AND GRADE OF OFFICER OR OFFICIAL ON DUTY | SIGNATURE

DA FORM 1 NOV 62 1594 CONFIDENTIAL REPLACES EDITION OF 1 NOV 62 WHICH IS OBSOLETE.

Radisa A

CONFIDENTIAL

DAILY STAFF JOURNAL OR DUTY OFFICER'S LOG (AR 220-345)				PAGE NO. 3	NO OF PAGES 17

ORGANIZATION OR INSTALLATION	LOCATION		PERIOD COVERED	
S-3 Section 1st Bn, 7th Arty	Di An, Vietnam XT9007		FROM	TO
		HOUR 0001	DATE 7 Sept 69	HOUR 2400	DATE 9 Sept 69

ITEM NO	TIME IN	OUT	INCIDENTS, MESSAGES, ORDERS, ETC.	ACTION TAKEN	INI-TIALS
(cont)	0258		Flareship running low — want to crank up on calls		
			884395 ^{B} 881390 ^{B}		
			887391 ^{B}		
			880396 ^{B}		
	0302		Major Samack gave permission to exceed ASR on 155 Illum. (DO) at Arc light		
	0303		50 E released at this time		
	0305		Impact on ^{B} 881386 for A8/6		
			I 890388 — Max Charge, may		
			do a lot of dmde – will have trouble keeping guns in the pits		
	2320		Further Sitrep A 30 reported		
			neg. mortar missing took out RPG i r grenades		
	2325		C8/6 I shot AH 9012 ^{B} 880396 I 877807		
	2320		C8/6 I shot shot		
	2321		Checking for another dustoff		
	2326		Checking lifted, C8/6 I shot again		
	2328		Shot, MGP on every 90 sec		
	2330		Be informed we Welders lost 1 KIA on FSB Jim		
	2335		C8/6 shooting streamers		
	2341		C8/6 I checkfire all flare ellipse WOP only us correcting		

TYPED NAME AND GRADE OF OFFICER OR OFFICIAL ON DUTY	SIGNATURE

CONFIDENTIAL

DA FORM 1594   PREVIOUS EDITIONS OF THIS FORM IS OBSOLETE.   PFC-Japan

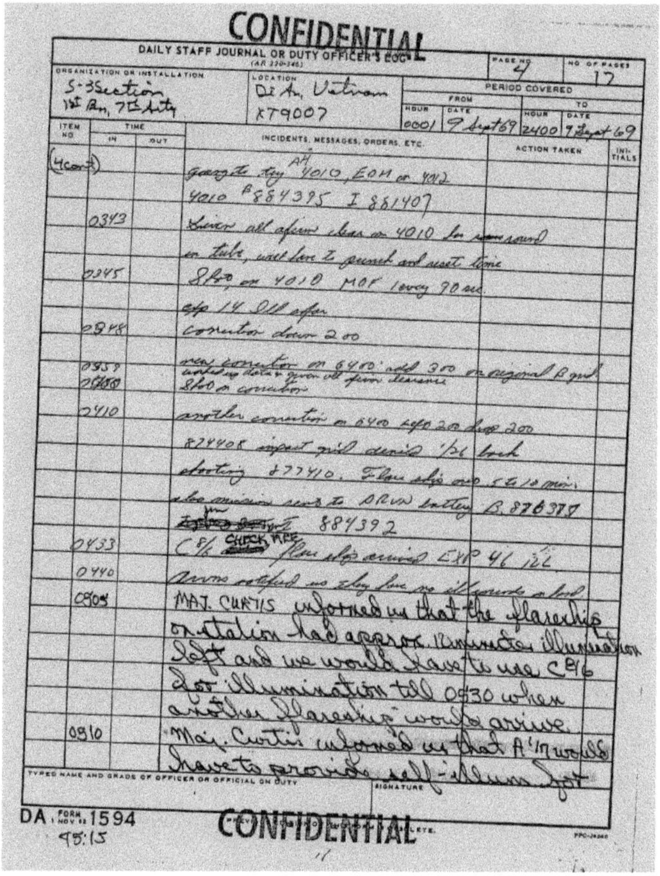

CONFIDENTIAL

DAILY STAFF JOURNAL OR DUTY OFFICER'S LOG (AR 220-346)

ORGANIZATION OR INSTALLATION	LOCATION	PAGE NO	NO OF PAGES
S-3 Section, 1st Bn, 7th Arty	Di An, Vietnam XT9007	4	17

PERIOD COVERED: FROM Hour 0001, Date 9 Sept 69; TO Hour 2400, Date 9 Sept 69

ITEM NO	TIME IN	TIME OUT	INCIDENTS, MESSAGES, ORDERS, ETC.	ACTION TAKEN	INITIALS
(4 cont)			[illegible] 4010, EOM [illegible] 4012		
			4010 B 884395 I 881407		
	0343		[illegible] 4010 [illegible]		
			[illegible]		
	0345		[illegible] 4010 MOF [illegible] 90 [illegible]		
			[illegible]		
	0348		correction down 200		
	0359		new correction on 6400 [illegible] 300 [illegible]		
	0400		[illegible]		
	0410		another correction on 6400 [illegible] 200 [illegible] 200		
			874408 [illegible]		
			[illegible] 877410. [illegible]		
			[illegible] ARVN battery B 876378		
			[illegible] 884392		
	0433		C 8/6 CHECK FIRE [illegible] EXP 46 [illegible]		
	0440		[illegible]		
	0905		MAJ. CURTIS informed us that the flareship on station had approx. 12 minutes illumination left and we would have to use C 8/6 for illumination till 0930 when another flareship would arrive.		
	0910		Maj. Curtis informed us that A 1/7 would have to provide self-illum. for		

TYPED NAME AND GRADE OF OFFICER OR OFFICIAL ON DUTY | SIGNATURE

DA FORM 1 NOV 62 1594

CONFIDENTIAL

CONFIDENTIAL

DAILY STAFF JOURNAL OR DUTY OFFICER'S LOG
(AR 220-346)

PAGE NO. 5 | NO. OF PAGES 17

ORGANIZATION OR INSTALLATION: S-3 Section, 1st Bn, 7th Arty
LOCATION: Di An, Vietnam XT9007
PERIOD COVERED: FROM HOUR 0001 DATE 9 Sept 69 — TO HOUR 2400 DATE 9 Sept 69

ITEM NO	TIME IN	TIME OUT	INCIDENTS, MESSAGES, ORDERS, ETC.	ACTION TAKEN	INITIALS
(4 cont)			five minutes while the spooky moves off station.		
	0520		A 1/7 providing self-illumination		
	0525		A 1/7 EOM EXPENDED 3 ILL		
	0526		C 8/6 BEGINNING to fire illumination on AH 9010		
	0528		A 1/7 requests pallet of sandbags in the morning		
	0540		Dabol informed us that estimated time of arrival for flareship was 0630. C 8/6 would have to provide Ill. till then		
	0600		Phuoc Vinh gave A 1/7 air clearance 4 clicks around and 5000 feet up.		
	0625		C 8/6 EOM on illumination mission Expended 100 ILL		
	0850		Dagger reports credit for 1 KBA ARTY for A 1/7 in Rat Patrol support		[illegible]
	late entry 0800		A 1/7 cancells air data around fire with Phuoc Vinh.		[illegible]
5	0900	(4)	Possum 65 7 reported that his possum people will be using their [illegible] for the next couple of days		[illegible]

TYPED NAME AND GRADE OF OFFICER OR OFFICIAL ON DUTY | SIGNATURE

CONFIDENTIAL

DA FORM 1 NOV 62 1594 — EDITION OF THIS FORM IS OBSOLETE.

CONFIDENTIAL

DAILY STAFF JOURNAL OR DUTY OFFICER'S LOG (AR 220-345)				PAGE NO. 6	NO. OF PAGES 17
ORGANIZATION OR INSTALLATION S-3 Section 1st Bn, 7th Arty	LOCATION Di An, Vietnam XT 9007			PERIOD COVERED	
		FROM 0001 7 Sept 69		TO 2400 7 Sept 69	

ITEM NO	TIME IN	OUT	INCIDENTS, MESSAGES, ORDERS, ETC.	ACTION TAKEN	INITIALS
6	1030	(W)	B1/2 reports they are having problems with their FADAC. They had previously thought that it was a power problem, they now think it to be a malfunction of the blower. Comms has been informed, and will take care of sending contact team out		
7	1055	(W)	F1/6 and B1/12 will register, F1/6 will register on 994262, B1/12 on 905251. Ground clear X8 (GRN) Tam Uyen (CLS) Air clear Bien Hoa (DC)		
	1100		F1/6 shot - 1 of 3 - splash 1 - shot 2 -		
	1101		splash 2 - shot 3 - splash 3		
	1102		F1/6 - L100 + 100		
	1103		F1/6 shot - splash - D50 - FFE		
	1104		F1/6 shot splash - over line 30 meters		
	1105		F1/6 shot 1 of 4 F1/6 shot 2 - splash 1		

DA FORM 1594

CONFIDENTIAL

154

CONFIDENTIAL

DAILY STAFF JOURNAL OR DUTY OFFICER'S LOG (AR 220-346)

ORGANIZATION OR INSTALLATION	LOCATION	PAGE NO	NO. OF PAGES
S-3 Section, 1st Bn, 7th Arty	Di An, Vietnam XT 9007	8	17

PERIOD COVERED			
FROM		TO	
HOUR	DATE	HOUR	DATE
0001	9 Sept 69	2400	9 Sept 69

ITEM NO	TIME IN	TIME OUT	INCIDENTS, MESSAGES, ORDERS, ETC	ACTION TAKEN	INITIALS
(9 cont)	1147	cont	C&C 8 remain with 1/7 ARTY until 1700 hrs		AJ
10	1147	u	50A setting down at Phuloi & repair radio.		AJ
11	1210	(u)	Metro Station Bien Hoa was contacted and are, because of equipment failures, unable to compute the 0800 release Met. A 2/13 has Cu Chi met and will pass it on to B 7/12		[illegible]
12	1300	(c)	LRRP AO Extension cancelled by Dagger, replaced by BH TAC AO extension Ter o/o: coord of Boundary; 000340 To 000280 (straight line), follow 1/26-1/18 boundary around To 140340, Then To 000380 (straight line).	Posted 1300.	AJ
13	1320	(u)	69 will be registered on DIL 905251 50A will be the observer [illegible] Passing clearance (DHM) air clearance Phu Loi Arty (NLB).		[illegible]

TYPED NAME AND GRADE OF OFFICER OR OFFICIAL ON DUTY — SIGNATURE

CONFIDENTIAL

DA FORM 1 NOV 62 1594 — PREVIOUS EDITION OF THIS FORM IS OBSOLETE.

FPC-Japan

CONFIDENTIAL

DAILY STAFF JOURNAL OR DUTY OFFICER'S LOG
(AR 220-346)

ORGANIZATION OR INSTALLATION	LOCATION	PAGE NO.	NO OF PAGES
S-3 Section 1st Bn, 7th Arty	Di An, Vietnam XT9007	9	17

PERIOD COVERED			
FROM		TO	
HOUR	DATE	HOUR	DATE
0001	9 SEPT 69	2400	9 Sept 69

ITEM NO.	TIME IN	TIME OUT	INCIDENTS, MESSAGES, ORDERS, ETC.	ACTION TAKEN	INITIALS
(13 cont)	1345		B2/12 shot- splash		[illegible]
	1346		B2/12 late shot 1, 2 : splash.		[illegible]
	1348		B2/12 L100, +200		A/
	1353		B2/12 shot, splash;	R50, +100	A/
	1357		B2/12 late shot, splash;	L25, -50 ffe	A/
	1400		~~B2/12~~ B2/12 shot, splash;	+ Ln	A/
	1400		B1/7 has complied with all new NFZ overlays; initials from B1/7	M.G.C.	A/
	1401		B2/12 shot, splash;	-R	A/
	1403		B2/12 shot 1, 2, 3 ; splash 1, 2, 3 ;	-Ln, -Ln, -Ln	A/
	1409		B2/12 shot, 1 splash 1;	+ Ln (Valid Reg).	A/
	1411		B2/12 shot, splash;	+ Ln, E.O.M.	A/
	1412		50A released 8 L.K.		A/
	1413		B2/12 expended 11 HE/Pd convey.		A/
	1430		Possum 65S Reports NO ONE a This location Gave the authority for Flash Light 111 To SP To Lai Khe. Initials DWH To Possum 65's Knowledge. Further SitRep To follow upon questioning Possum 65Y and 65U.		A/
14	1500	(U)	We informed Dagger and Champion that 1-Company of infantry and 1 battery A 2/32 will displace		A/

TYPED NAME AND GRADE OF OFFICER OR OFFICIAL ON DUTY | SIGNATURE

DA FORM 1 NOV 62 1594 — CONFIDENTIAL — [illegible] OBSOLETE.

DAILY STAFF JOURNAL OR DUTY OFFICER'S LOG — CONFIDENTIAL

ORGANIZATION OR INSTALLATION	LOCATION	PERIOD COVERED	PAGE NO. 10	NO. OF PAGES 17

Organization: 3-3 Section, 1st Bn, 7th Arty
Location: Di An, Vietnam — XT9007
From: 0001 9 Sep 69 — To: 2400 9 Sep 69

ITEM NO.	TIME IN	OUT	INCIDENTS, MESSAGES, ORDERS, ETC.	ACTION TAKEN	INITIALS
(cont)			to our A.O. and come under our control. They are 82nd Abn units.		
15	1522	(u)	Cords 30 Landing F 1/7 Location.	30	
16	1600	(u)	Cpt Bivens a Pagoda Tan built with the Radar Callsign 64 F Also Radar Sta in AO are operational		
17	1600	(u)	A'ty is on Do Not Load Missions Grid 910 445 for Wilder 86 Negative Setup at this time		
17	1610	(u)	To all FSB, Base Camps, and NDPs: Inspect perimeter wire and insure wire is complete and in place Inspect Trip flares and insure these in place All drainage deflected to prevent infiltration All TOC's must be checked Ensure security		

DA FORM 1594 — CONFIDENTIAL

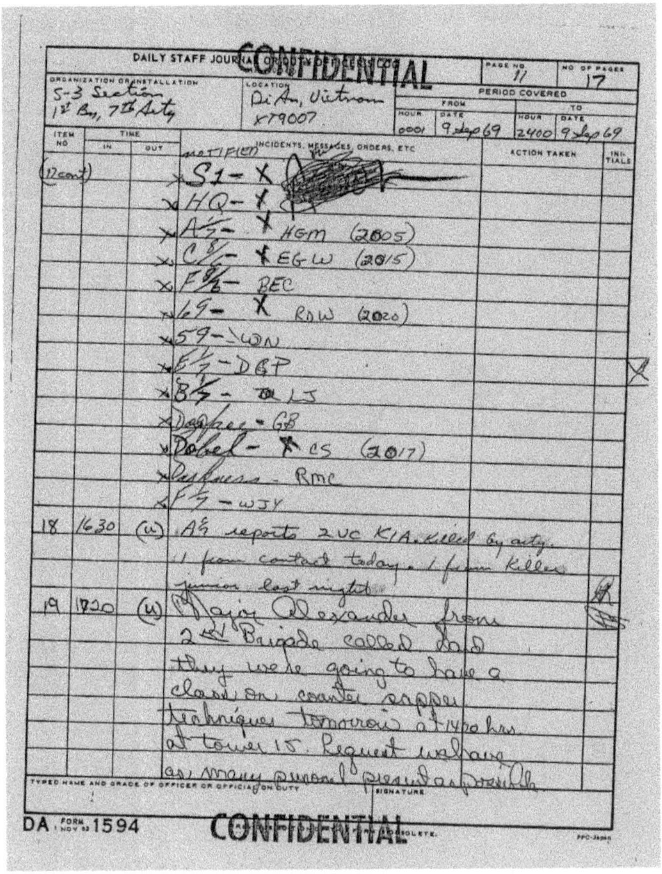

CONFIDENTIAL

DAILY STAFF JOURNAL OR DUTY OFFICER'S LOG

ORGANIZATION OR INSTALLATION	LOCATION	PAGE NO	NO OF PAGES
S-3 Section, 1st Bn, 7th Arty	Di An, Vietnam XT9007	11	17

PERIOD COVERED FROM: HOUR	DATE	TO: HOUR	DATE
0001	9 Sep 69	2400	9 Sep 69

ITEM NO	TIME IN	TIME OUT	INCIDENTS, MESSAGES, ORDERS, ETC	ACTION TAKEN	INITIALS
(17 cont)			notified		
			x S1 – X [illegible]		
			x HQ – X		
			x A 1/7 – X HGM (2005)		
			x C 8/6 – X EGW (2015)		
			x F 8/6 – BEC		
			x 69 – X RDW (2020)		
			x 59 – WDN		
			x E 1/7 – DGP		
			x B 1/7 – LJ		
			x Dagger – GB		
			x Dobol – X CS (2017)		
			x Dusters – RMC		
			x F 1/7 – WJY		
18	1630	(W)	A 1/7 reports 2 VC KIA. Killed by arty. 1 from contact today. 1 from Killer junior last night		
19	1720	(W)	Major Alexander from 2nd Brigade called said they were going to have a class on counter sapper techniques tomorrow at 1400 hrs. at tower 15. Request we have as many personnel present as possible		

TYPED NAME AND GRADE OF OFFICER OR OFFICIAL ON DUTY | SIGNATURE

DA FORM 1 NOV 62 1594 — CONFIDENTIAL

DAILY STAFF JOURNAL OR DUTY OFFICER'S LOG
(AR 220-345)

CONFIDENTIAL

PAGE NO. 12 NO OF PAGES 17

ORGANIZATION OR INSTALLATION	LOCATION	PERIOD COVERED			
S-3 Section 1st An, 7th Arty	Di An, Vietnam X79007	FROM HOUR 0001 DATE 9 Sep 69	TO HOUR 2400 DATE 9 Sep 69		

ITEM NO.	TIME IN	TIME OUT	INCIDENTS, MESSAGES, ORDERS, ETC.	ACTION TAKEN	INITIALS
20	1730	(u)	Dagger 24K reports the Brigade school has cancelled their operations for tonight.		PS
21	1740	(u)	64 E has closed Pagoda Inn and will be operational in approx. 30 min		GR
22	1836	(u)	Yellow Jacket 086319 early afternoon. No Fire.		GR
23	1830	(c)	FCL N40 S50 E87 W94 STRIKE GRID 3843 - Called down by 65 F		PS
24	1841	(u)	64E is operational at this time at PI		OP
	1842		Searchlight has been moved from PI.		
25	1852	(u)	7CC from Dubol lifted		CP
20	1900	(u)	Cpt Jarina designated that only 1 guard would be used around the TOC with the assumption that one gate would be locked during the hours of darkness. This will be procedure unless otherwise changed by ___		

TYPED NAME AND GRADE OF OFFICER OR OFFICIAL ON DUTY	SIGNATURE

DA FORM 1594
NOV 62

CONFIDENTIAL

CONFIDENTIAL

DAILY STAFF JOURNAL OR DUTY OFFICER'S LOG
(AR 220-346)

PAGE NO 13 | NO OF PAGES 17

ORGANIZATION OR INSTALLATION: S-3 Section, 1st Bn, 7th Arty
LOCATION: Di An, Vietnam XT9007
PERIOD COVERED: FROM HOUR 0001 DATE 9 Sep 69 — TO HOUR 2400 DATE 9 Sep 69

ITEM NO	TIME IN	TIME OUT	INCIDENTS, MESSAGES, ORDERS, ETC.	ACTION TAKEN	INITIALS
(20 cont)			the S-3, Bn XO, or Bn Co.		JRB
21	1907	(C)	PI RPTS RPG FIRE AT 860210 — NEGATIVE ACTION TAKEN AT This Time		JRB
22	1913	(U)	OA reports guards on station or Bikem.		JRB
23	1925	(U)	Div G-3 had requirement for flashlight from PI at Lai Khe, assignment for flashlight at PI terminated when US elements vacated PI.		JRB
24	1945	(C)	GIDDY PROFIT — called IN A CONTACT AT 823091 — REQUEST U.S. To STANDBY		CK
25	1947	(C)	VN little Saver Base 833121 — Sneak 90 element Per Bon		
26	2020	(U)	Cpt James informed Dunn fire security check has been completed the station		JRB

TYPED NAME AND GRADE OF OFFICER OR OFFICIAL ON DUTY | SIGNATURE

DA FORM 1 NOV 62 1594 — EDITION OF ... THIS FORM IS OBSOLETE.

CONFIDENTIAL

DAILY STAFF JOURNAL OR DUTY OFFICER'S LOG

CONFIDENTIAL

ITEM NO.	TIME IN	TIME OUT	INCIDENTS, MESSAGES, ORDERS, ETC.	ACTION TAKEN	INITIALS
36	2344	(w)	A¾ EOM on ill to N		
	2355		A¾ EOM on ill to S. negative findings will sweep again at first light of morning. A¾ exp. 85 ill.		
37	2158	(u)	E¼ in SHOT		
38	2400	(c)	Summary: OP TOAN THANG, PHASE III continued with 1st Bn, 7th Arty providing supporting fires for 2nd Bde and on Call Fires to adjacent FW MAF and ARVN units. Organization for combat is as follows: A/1/7 (105) FSB Jim (XT 848 391), B/1/7 (105) FSB Marietta (XT 702 145), C/1/7 (105) FSB Thunder III (XT 771 695) supporting 1st Bde operations. E/1/7 (4.2) Di An, F/1/7 (4.2) [Phu Loi] OPCON 82d Abn. to support 3rd Bde, PH (+) B/7/7 192 (155) GSR ½ Arty (Phu Loi), C/8/6 Arty (155) GSR ½ Arty III XT 907 319), 2 Sections 1st Plt, Btry, 29th Arty, attached ½ Arty (¾ Section Di An, ½ Section Phu Cuong Bridge XT 801 144, ½ Section Pagoda bun XT 873 186, ½ Section Aa Byn Bridge XT 781 132).		

TYPED NAME AND GRADE OF OFFICER OR OFFICIAL ON DUTY   SIGNATURE

DA FORM NOV 59 1594   CONFIDENTIAL

CONFIDENTIAL

DAILY STAFF JOURNAL OR DUTY OFFICER'S LOG
(AR 220-345)

ORGANIZATION OR INSTALLATION	LOCATION	PERIOD COVERED			
S-3 Section 1st Bn, 7th Arty	Di An, Vietnam XT9007	PAGE NO. 17	NO. OF PAGES 17		

PAGE NO. 17 NO. OF PAGES 17

FROM: HOUR 0001 DATE 9Sep69   TO: HOUR 2400 DATE 9Sep69

ITEM NO	TIME IN	TIME OUT	INCIDENTS, MESSAGES, ORDERS, ETC.	ACTION TAKEN	INITIALS
(38cont)			Targets engaged during the period 0001-2400 9 September 1969 were as follows: Wake up round -1, Contact -1, Movement -1. Ammunition expended upon these targets was as follows: Wake up - 1(175) HE, Contact-16(175) HE, PD, Movement- 85(105) HE.		
39	2400		(u) Journal Closed		

TYPED NAME AND GRADE OF OFFICER OR OFFICIAL ON DUTY
STEPHEN G. JOHNSON, Major, FA, S-3   SIGNATURE

DA FORM 1594, 1 NOV 62

CONFIDENTIAL

Appendix C; Award of the Vietnam Cross of Gallantry with Palm

6. The Cross of Gallantry with Palm has been awarded by the Government of the Republic of Vietnam for outstanding service during the period 15 September 1969 through February 1970 to:

1ST BATTALION, 7TH ARTILLERY (1ST INFANTRY DIVISION)

The citation reads as follows:

The 1st Battalion, 7th United States Artillery, 1st Infantry Division, is an outstanding United States Artillery Unit with great combat ability and outstanding fire support techniques. Operating in Vietnam from July 1965 to April 1970, tested by repeated clashes with the Vietnamese Communists, the men of the 1st Battalion, 7th United States Artillery, have always set a gallant example before the unit, with their will to fight and win, their acceptance of every sacrifice together with their appreciation of methods of troop maneuver and modern fire techniques for use against the enemy. The 1st Battalion, 7th United States Artillery, 1st Infantry Division, successively and outstandingly defeated several units of the Communist aggressors, at the same time creating favorable conditions for Infantry to clear out enemy rear bases and barracks in secret zones within the area of tactical responsibility of the 1st United States Infantry Division. In particular, in September 1969, during Operation 'TOAN THANG' (Phase III), the 1st Battalion, 7th United States Artillery, was moved to encircle the enemy in PHU HOA DONG Hamlet and assist the 1st Battalion, 18th Infantry, in winding up the battle. During 8 successive days of siege, before any assaults were launched the gallant men of the 1st Battalion, 7th Artillery, fired into the enemy's hiding places, creating an intense artillery fire fence and foiling all the Communists' attempts to escape from the envelopment, leveling many of their combat fortifications and weakening their fighting potential and spirit. After that, the men of the 1st Battalion, 7th Artillery, assisted the 1st Battalion, 18th Infantry, in blitzkrieg assaults to seize the objective. As a result, the 1st Battalion, 7th Artillery, accounted for many of the enemy killed in action and a number of others buried in their trenches. During Operation 'TOAN THANG' (Phase III and IV), except for engagements with major Communist units, the batteries of the 1st Battalion, 7th Artillery, were usually dispersed throughout the operating area of the 1st Infantry Division. Military operations full of dangers and hardships never deterred the men of the 1st Battalion, 7th Artillery, from fighting. The rounds fired from their batteries inflicted many personnel casualties and much damage to the food supplies and military equipment of the Communist aggressors and paralyzed their combat potential and hopes of launching a new general offensive in the III Corps Tactical Zone, especially during the period before the 1970 Vietnamese New Year. With the above combat record and the numerous splendid victories achieved, the 1st Battalion, 7th United States Artillery, 1st Infantry Division, has materially contributed to the fight against Communist aggression in the Republic of Vietnam.

The Vietnam Cross of Gallantry with Palm was awarded during the author's tour of duty.[2]

---

[2] The 1/7th Artillery also received the Vietnamese Civil Action unit citation.

Appendix D; Locations and Coordinates of Fire Support Bases Associated with the First Infantry Division.

FSB Aachen (XT625503) Dec 68

An Loc MACV Compound (XT7688)

Ap Bo La (XT8839)

APF Compound (XS963801)

FSB Apollo (XT637507) October 69

FSB Aspen (XT747801) August 1969

Bien Hoa Airbase (YT041172)

Binh Ninh (XT721858) 214 RF Company July 1969

Bunard (YU266884)

FSB Buttons (YU139073)

FSB CHARLIE (XT5685) (JUNCTION CITY II)

FSB Christmas (XT833522) September 69

FSB Dogpatch (XT620645) 1969

FSB Drac (XT903610) Dec 68

Duc Vinh III (7580) (SF An Loc)

Duster Compound (YT127117)

FSB Freda (XT583333) July 69 (Later renamed FSB Tennessee)

FSB Gela (XT604432/605420) September 69

FSB Greene (XT642708) May 1969

FSB GUNNER I (XT995545) (JUNE 67)

FSB Harpers Ferry (XT908306)/(XT912302) Dec 68

FSB Howard (XT703817)

LZ Jamie (XT487721) 1969

LZ Joe (XT628664) 1969

FSB Julie (XT5980) 28 Oct-10 Nov 68

Junction City (XT610348)

FSB Kien (XT521415) September 69 (formerly FSB Mahone)

Lai Khe Basecamp (XT770380)

FSB Lorraine (XT712409) September 69

Loc Ninh SF/CIDG (XU8213)

Loc Ninh SF/CSF (XU733083)

Lundy's Lane (XT575502) July 69

MACV Compound U/I (YU155078)

FSB Mons V (XT583344)

FSB Mons XII (XT636347) September 69

FSB Normandy (XT90273190) July/August 69

FSB Omaha (XT890547) Dec 68

FSB Oran (XT628509)/(XT631504) Dec 68

LZ Phyllis (XT532816) 1969

FSB Pine Ridge (XT522587) October 69

Quan Loi Basecamp (XT815904)

FSB Ramrod (624572) July 69

Remagen I (YT005565)

Remagen II (YT002612)

Remagen III (YT032633)

Remagen IV ((YT056675)/(YT005675)

Remagen V (XT061719)

Remgen VI (YT138792)

Remagen VIII

FSB Rawlings (XT297486)

FSB St. Barbara (XT279679) 1969

FSB Sam (XT592610) May 1969

FSB Sidewinder (XT739821) August 1969

FSB Son (XT682485) September 69

Song Be (YU155078)

Tan Binh East (XT856360)

Tan Binh West (XT876364)

Tan Hung (XT 863876) 399 RF Company August 1969

FSB Tennessee (XT583333) October 69

FSB Thunder II (XT78175560) September 69

FSB Thunder III (XT772656)/(XT772646) Dec 68

FSB Thunder IV (XT763893)

Tong Le Chon (XT624875)

Bibliography

Kelley, Michael P. 2002. *Where We Were in Vietnam: A Comprehensive Guide to the Firebases, Military Installations and Naval Vessels of the Vietnam War, 1945-1975* Midpoint Trade Books.

Murphy, Edward F. 2007 *Dak To, America's Sky Soldiers In South Vietnam's Central Highlands* Ballantine Books, New York.

Stanton, Shelby L. 1986 *Vietnam Order of Battle* Galahad Books, New York.

Unpublished *Annual Historical Summary of the 1st Battalion, 7th Artillery for the period January 1, 1969 to March 19, 1970* U.S. National Archives and Records Administration, College Park, Maryland.

Walt Cross is a retired U.S. Army master sergeant. He is a veteran of the Vietnam War serving with Battery A, 1st Battalion, 7th Artillery, 1st Infantry Division from 1968 to 1970. He holds university degrees in both history and science. He is a graduate of the U.S. Army Sergeants Major Academy. Among his decorations are the Bronze Star with "V" device for heroism in combat with an Oak Leaf Cluster, Meritorious Service Medal, Army Commendation Medal with 2 Oak Leaf Clusters, Army Achievement Meal with Oak Leaf Cluster, and various service and campaign medals. He resides in Stillwater, Oklahoma.

www.ingramcontent.com/pod-product-compliance
Lightning Source LLC
Chambersburg PA
CBHW070837100426
42813CB00003B/654